Bull Brook,
Massachusetts

Chesapeake Bay,
Virginia

Meadowcroft Rockshelter,
Pennsylvania

Knife Lake,
Minnesota

Cactus Hill site,
Virginia

Anzick site,
Montana

La Sena site,
Nebraska

Big Bone Lick,
Kentucky

Lovewell site,
Kansas

Hell Gap,
Wyoming

Lindenmeier site,
Colorado

Tennessee River Basin,
Tennessee

Topper site,
South Carolina

Snowmass,
Colorado

Mahaffy Cache,
Colorado

Aucilla River,
Florida

Lime Ridge site,
Utah

Blackwater Draw,
New Mexico

Bechan Cave,
Utah

Buttermilk Creek, Texas

Trinity site,
New Mexico

Mockingbird Gap site,
New Mexico

Hoyo Negro,
Quintana Roo

Lehner Mammoth Kill site,
Arizona

El Fin del Mundo,
Sonora

Cerutti Mastodon site,
California

ALSO BY CRAIG CHILDS

Apocalyptic Planet
Finders Keepers
The Animal Dialogues
House of Rain
The Way Out
Soul of Nowhere
The Secret Knowledge of Water

ATLAS OF A LOST WORLD

ATLAS OF A LOST WORLD

Travels in Ice Age America

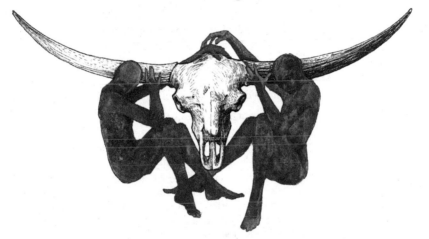

CRAIG CHILDS

Illustrations by Sarah Gilman

PANTHEON BOOKS · NEW YORK

Library of Congress Cataloging-in-Publication Data
Name: Childs, Craig, [date] author.
Title: Atlas of a lost world : travels in ice age America / Craig Childs.
Description: New York : Pantheon, 2018.
Identifiers: LCCN 2017033037. ISBN 9780307908650 (hardcover).
ISBN 9780307908667 (ebook).
Subjects: LCSH: Prehistoric peoples–North America. Paleo-Indians–North America.
Glacial epoch–North America. Paleoecology–North America–Pleistocene.
Mammals, Fossil–North America. BISAC: HISTORY/North America.
SCIENCE/Paleontology. NATURE/Ecology.
Classification: LCC E77.9 .C55 2018. DDC 551.7/92–dc23.
LC record available at lccn.loc.gov/2017033037

www.pantheonbooks.com

Jacket design by Tyler Comrie
Book design by Iris Weinstein
Endpaper map copyright © 2018 by David Lindroth Inc.

Printed in the United States of America
First Edition
4 6 8 9 7 5 3

For Sharon, the mom who whispered in my ear, Go

It's how they say
we arrived here, from a cave
in the beginning.

—LINDA HOGAN

CONTENTS

PROLOGUE

Nothing stirred in the bowels of the earth. Bones lay on bones in silence, a cave sealed shut, its entrance collapsed in the middle of the Pleistocene, the last geologic epoch before our own. Anything stuck inside never got out. The first inkling of new life was the white light of a headlamp and the scratch of small metal tools.

I lay on my side with my hard hat off, working on a camel that had died back here, an awkward place for anything to die, I thought. It was in a crack in the back side of a small chamber, which is why they gave the site to me. I was in my early thirties, base camp cook for a museum crew working on one of the most prolific Pleistocene bone-producing sites in the American West—Porcupine Cave in the mountains of south-central Colorado—churning out rodent bones for the most part, with the occasional *hoorah* of discovering skeletal cheetah kittens, the toe bones of horses, cores of bison horns. Elsewhere, a dozen excavators and screeners were pulling up a treasure trove of Pleistocene porcupines and squirrels, filling and labeling bags, while I was alone in a hole with *Camelops hesternus*. I imagined the camel crawling into the cave, its hindquarters slashed by saber-tooths or hamstrung by dire wolves, trying to save itself, dying here.

When silver miners first blasted into the mountainside in the 1860s, they accidentally clipped one of the cave's chambers. So they propped up the tunnel with wooden posts and beams. Then they dropped into it with their oil-wick lamps and found passageways, crystals, and domed rooms—floors buried in dust, bones sealed in place three hundred thousand years ago. Now, a ladder brought paleontologists down through the hole the miners opened. They moved through the rotten molasses smell of woodrat urine into the dark, crawling through low spots, pushing gear ahead, headlamps shining on each other's boot soles. They went to their own trenches and pits,

screens and vacuums, setting up grids, fingers picking through countless rodent bones. I went to the camel.

Animal bones had been dragged in by woodrats, though some
skeletons, like that of the camel, were complete. The biggest mammals of the age were being found elsewhere, rock shelters and muddy
pond bottoms holding the remains of mammoths, mastodons, giant
bison, dire wolves, large fanged cats, sloths with foot-long claws, and
a six-foot-tall beaver with incisors like medieval weapons. These creatures, including the camel, are known as *Rancholabrean megafauna,*
named after copious Ice Age finds at the La Brea Tar Pits in Southern
California, some of the largest mammals of the Pleistocene.

Gigantism among mammals is associated with cold. This is known
as Bergmann's Rule, after the nineteenth-century German biologist
Carl Bergmann, who noticed that the larger examples of most species
tend to occupy colder climates, while the smaller tend to be found
in warmer places. Warmth produces smaller bodies that expel heat,
while cold encourages extra layers of big bones, muscle, fat, and fur,
resulting in bodies with a smaller ratio of surface area to volume and
a greater ability to hold in heat. Glacial periods—long cold spells that
often lasted one hundred thousand years at a time—were frequent in
the Pleistocene, when the earth on average dropped 10 to 15 degrees
Fahrenheit below temperatures we experience today. This gave rise to
animals with large bodies and Ice Age megafauna weighing a ton or
more. The largest American mammoths came in at around ten tons,
three tons more than the largest African elephants.

I was drawing slow circles around the eye of an animal that had
once weighed about seventeen hundred pounds, just shy of a ton. I
picked out pieces of yellowish earth from where one of its eyeballs
had once rotated in its socket, seeing big cats coming, watching
sunsets on glaciers and wind blowing across the lush intermountain
grasslands.

None of the animals found in this cave had encountered a human
being. The scientists joked about finding an artifact, a stone projectile, but it wouldn't happen. This camel never knew spears or hunting dogs, or the smell of campfires. The land from Alaska to the tip
of South America—nine thousand latitudinal miles of mountains,
rivers, plains, and coastlines—had no humans, no deliberately sharpened rocks or bones flaked and polished to a point. This was the last
large habitable part of the world that the human race discovered,

nobody arriving until the tail end of what is known as the Wisconsin Ice Age, a glacial period that lasted about one hundred thousand years and ended ten thousand years ago at the close of the Pleistocene. Neighboring Eurasia had been foraged by tool-wielding hominids for a million years or more. Early human species from Spain to Indonesia were cutting meat and skins with stone knives, building fires, sometimes burying their dead in caves and rockshelters, and duking it out with other hominids. All the while, nobody made it this far around the globe.

The first humans, anatomically modern *Homo sapiens*, left Africa between one hundred thousand and seventy thousand years ago. These people were nearly identical to us. Give them a haircut and modern clothes, and they'd be hard to pick out in a crowd, their size, hairiness, and skin color easily within our own contemporary range. They left their home continent along the bottom of Saudi Arabia, jumping the Red Sea at the strait of Bab el-Mandeb at Djibouti, the same place baboon populations had already swum across to populate the Middle East. It is as if humans saw the baboons and dived in after them, figuring they had to be going somewhere.

From there, the path of humans is marked by stone blanks called *preforms*. They appear to have been an invention of *Homo sapiens*, a stone worked down to carrying size, something that could be fashioned into whatever might be needed: blades, projectiles, scrapers, or the small hand-chisels called *burins*, pointed rocks gripped between fingers and thumb and used to sharpen and refinish weapons and tools. Instead of grabbing whatever rock could be found and working it into a sharp wedge, as early hominids were wont to do, this new people had the best resources in hand and ready to go. Their traveling tool kits allowed them farther and faster movement, and access to better weapons and tools as needed.

A weapon called a *biface* spread along with these preforms. Bifaces were stone points and blades knapped from both sides for lightness and efficiency. Our species was armed, mobile, and pushing rapidly into the world, going beyond the reach of anything before them, and the world kept opening to them.

By forty-eight thousand years ago, humans reached the unpopulated Pleistocene super-continent of *Sahul*, which consisted of Australia, Tasmania, and New Guinea fused together at lower sea levels. It is assumed that people used boats to get there. The minimum number of people needed for even a remote chance of long-term survival is forty, meaning at least that many had to cross open water to reach Sahul. They were not a desperate handful clinging to a log, adrift and without a plan, but a group making a concerted effort, a small and daring population deliberately taking to the water, convinced something worthwhile was out there.

Around the same time, other populations of *Homo sapiens* were pushing beyond their northern boundaries into lobes of ice and the mammoth-studded steppe of upper Siberia. A mammoth butchered by stone or bone tools forty-five thousand years ago was found in the Siberian North near the Arctic Sea. A forty-thousand-year-old human leg bone emerged from the banks of the Irtysh River in western Siberia, its DNA half human and half Neanderthal.

There was never an easy way to get to this side of the planet. Hemmed in by the world's two largest oceans, the Pacific and Atlantic, and often capped with ice, the Americas seemed to be off-limits. The only land connection was exposed during glacial periods when an Arctic land bridge appeared between Siberia and Alaska. By twenty-four thousand years ago, there were game-hunting settlements within a season's walk of the Bering land bridge, though some believe people made it across by forty thousand years ago, or even earlier.

Scientists squabble over the locations and dates of human arrival in the New World. They shy away from words like *earliest* or *first*, knowing there will always be new and earlier discoveries. What is evidence, what is not? Is that a bone point or just a pointed bone? Were those mammoth and mastodon remains deliberately broken by tools and butchered, or is the damage a natural result of erosion and trampling and time? Age makes the Paleolithic record hard to read. There is no agreed-upon entry point or date. Artifacts were not left carpeting the ground, like souvenirs of a Moorish invasion. This was the farthest end of the world, the back door on human expansion. Numbers would have been few, and encampments would have been far between as the people found their way across the land bridge, kicking off one of the greatest experiments in global history.

This is a love story—boy meets girl, if you will. One partner is an unpeopled hemisphere, the other is our hungry, inquisitive species. Some might tell you that the encounter wasn't love at all, but domination, overkill, an invasive species hell-bent on spreading into a land that was doing just fine as it was, without us. Some scientists have called it blitzkrieg, mammoths felled like cordwood. Ours was no docile species, and the animals were not ready for us, or our weaponry. Archaeologists from Alaska to Florida have found Paleolithic spearpoints and stone blades still holding protein signatures from the meat of horse, camel, sloth, bison, bear, and the proboscideans, mammoths and mastodons. Ice Age bones in the Americas have been found scribed with human butchering marks, blackened from fires. But humans didn't always win. Many died; some were eaten. First people, wildly outnumbered by animals, would have found themselves tossed and trampled by tusks and hooves or torn to pieces by the scissoring teeth of scimitar cats. No matter how well-armed they were, even with Eurasian wolf dogs at their sides, surviving among Rancholabrean megafauna would have been challenging. Nobody said love would be easy.

I apologized out loud when I accidentally peeled a sliver of damp bone from the camel's skull. It seemed only polite to say something. The animal said nothing in return, silent as always. Our engagement back here felt like a first encounter between human and American megafauna. By sleight of scientific hand, I had slipped in before the first people, and this camel would have been bewildered to see me; it would never have encountered an animal so pale or so oddly limbed, gathered at its wildly bulging eye like a fleshy crab.

At lunch, everyone else left the cave and shut down the generator up top so they could eat in peace. I stayed below because my days here had to be quick; I had to leave the cave early, run two miles back to camp to start dinner and get a campfire going. With the generator off, all the work chambers fell into darkness as I kept working, the only light in the pit of the earth. I heard the *thunk* of a mammoth tusk striking rock on the surface above, where it swept through snow with its tusks, looking for winter browse. A Columbian mammoth, the largest of its kind in the world, was thirteen feet tall at the shoulder. I kept digging and scratching as I felt the distant boom of its steps as it moved its tonnage forward. Big cats would have been quieter, and I strained to listen for them padding across the ground above me.

I did not imagine the sound of people, no crack of stone on stone, or swish of a wooden rod making fire. Instead, I imagined the country empty of us, long before human eyes. In my imagination I rose up through the original entrance of Porcupine Cave, shielding my eyes from sunlight across glaciers, shaggy beasts of tusks and hooves grazing in the valley below. Continents would have been in the same positions, mountains and plains mostly unmoved, rivers in more or less the same places as they are now. More ice existed than we have today, storing water on land in the form of glaciers and ice caps, causing sea levels to drain by a few hundred feet, rewriting the coastlines.

Ten years ago is not hard to imagine. A hundred is well within grasp—you've probably known people who lived that long. A thousand years is ten lifetimes. Ten thousand years is one hundred lifetimes. It was around two hundred and fifty lifetimes ago that people began to arrive in North America into the teeth of Ice Age megafauna. I listened through the cave for the moment before this arrival, a time without time.

Under the pressing weight of limestone, no sign of light but my own, it became hard to tell which time was which. The skeleton was articulated, the animal had not moved from where it fell. Bones were in place like a time machine, the camel's head slumped into a low passageway. I was here like a seed blown into a cave, random, unexpected. How long would I last? How, without an atlas, would I find my place?

I listened for the Pleistocene, but all I heard was the *scratch scratch* of my fine metal pick, followed by the hush of the artist's brush I was using. In an ancient hole, only my headlamp shining through the blackness, I inhabited two worlds at once. One was an occupied continent streaming with freeways and lights. The other was this camel's land, a kingdom of animals, plants, and fungi. Eurasia would soon tip into the Americas. A land bridge with a navigable coast would be found.

I whispered to this camel, *Your world is about to change . . .*

ATLAS OF A LOST WORLD

1

LAND BRIDGE

DATE UNKNOWN

Before maps, we had elevation. We used mind instead of paper, and the contours of the land were recorded in our bodies. A knoll, a mountaintop, any high point would do. The earth unfolded before us.

From the flats of the Bering land bridge, the eye sees about a mile. Several hundred feet of elevation up a mountain or high hill can add thirty miles of visibility. You could have walked for weeks across the land bridge, through a mosaic of rivers and grasslands with hardly a rise in the ground, before you reached a small mountain range, isolated in the middle of it all. It was a cluster of gray volcanoes, standing like a beacon. No one would have known what lay ahead. The view from the flanks of these volcanoes would have revealed more of the same, no coast in sight, horizons continuing on as if constantly giving birth to themselves—mammoths, Pleistocene horses, and giant bears strung out as far as the eye could see. It must have seemed as if there were no end, the generosity of this planet unimaginable.

Around midnight I came up the slope of the Kookooligit Mountains. A storm was washing in from the north, clouds running low and wet across St. Lawrence Island in the Bering Sea, once the highest point on the land bridge.

Three degrees below the Arctic Circle in July, the hour was a luminous dusk, plum-colored storm touching ground, fingers of fog and cloud obscuring the distances. A little birthmark on the globe,

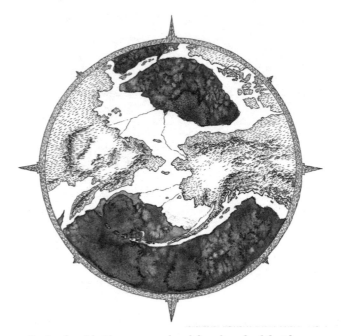

Bering land bridge at a sea level four hundred feet lower
than today, St. Lawrence Island situated just below the
almost touching thumbs of Siberia and Alaska

a fulcrum between continents, this island belongs to Alaska and the
United States and is occupied by Siberian Yup'ik people, as if the two
sides of the world could not help fusing here. St. Lawrence Island is
a remnant of the land bridge that once held Asia to North America,
now a misshapen bean ninety miles from end to end.

The name "Bering land bridge" is a misnomer. Never mind Vitus
Bering, the jowly Danish cartographer and explorer who led sailing
expeditions for the Russian Navy along the upper Pacific Rim in the
eighteenth century and after whom the Bering Sea and the Bering
Strait were named. When this was land and not water, the land bridge
was not a catwalk teetering from one hemisphere to the next, but a
flat subcontinent fully exposed when sea levels were at their glacial
low, its center five hundred miles from the nearest coast. I would have
looked across a steppe grassland and the occasional birch and black
spruce grove, summers free of snow, ground grazed and turned to
grass by large herbivores, loess blown in from the edges of distant ice
caps, allowing the soil to hold and retain organic matter. This would

have been an easily habitable landscape. Winters were dark and furiously cold, but summers produced copious wildflowers, their pollen found in cores taken from the bottom of the Bering Sea. The land bridge had experienced a unique regime of global weather patterns, the Pacific curling up warmly against its southern coast, Himalayan ice cap blocking precipitation from a quarter of the world away, and the mass of the land bridge itself holding its own temperature, a terrestrial heat sink. Inland precipitation was sparse, winter snows frigid but light. Land bridge summers were sunnier than those experienced on St. Lawrence Island, temperatures slightly warmer, more muskeg and grass than permafrost, snowpack melting earlier for longer growing seasons. This was the American Atlantis, and it went under wave by wave, storm by storm.

The sea has closed in, the ice caps melted, sea levels at a highstand and rising. This high point wouldn't have smelled of water-fat clouds drawing off a cold, dark sea. Every night, I took these walks, feeling the land bridge roll under my feet. I'd take advantage of the strange, intoxicating light and let go of our current geologic age, the Holocene, giving up my own time for another. The Mammoth Steppe of the Pleistocene would have extended away from me in great parks of grass and tundra, mammoths grazing, lions watching from the hill slope, one day never quite giving way to the next, as the sun touched just below the horizon before rising again. Some nights, I went out into sleet and I put my back against the wind, raincoat soaking through. Other nights, the sky lay across the island like a blanket, clouds insulating, and I moved in stillness. Tonight, windows of visibility opened between squalls, corridors of tundra revealed and removed like a hall of mirrors.

My mom had come with me on this trip, joining me on the plane out of Nome. Each night I waited till she fell asleep on her cot back in Savoonga, then slipped out the door. I was still carrying part of a sandwich she'd made but hadn't eaten, a snack for an evening stroll.

The first people to come here were not colonizers in the sense of a planned invasion, kings and queens lifting their chalices to fleets of ships bound for the horizon. If Ice Age people were sent by anything, it was the push and pull of the climate, the drive of human curiosity, and the tidal movement of other animals within which the people flowed. I doubt they would have considered themselves colonists, or that they had any idea of the scale of emptiness that lay ahead.

The sense of animals around these people would have been constant, woolly mammoths the size of garbage trucks, their tusks thirteen feet long and curved upward. Giant mammals were hungry, their homeothermic bodies requiring constant feeding in a vast grassland ecosystem at its evolutionary peak during the Ice Age. A healthy mammoth would have taken in more than six hundred pounds of browse a day and put out four hundred pounds of dung. Scientists working on a four-hundred-square-mile preserve of Pleistocene ecosystems in Siberia have analyzed the fossils found there to calculate original animal biomass, arriving at ten tons of animal per square kilometer. That's one mammoth, five bison, six horses, and ten caribou, plus an accompaniment of big cats and wolves. It wasn't necessarily emptiness ahead. The land was not abandoned. It was alive and had been so for a long time.

I'd been told not to walk more than three miles past the village. Any farther, and the land turns wilder. Not Ice Age wilder, just Holocene, small stuff. Polar bears have been sighted in the area, though the locals seemed more worried about wolves. I was told they descend from the mountains on stormy nights like this, using the fog as cover. The wolves sounded prehistoric in the descriptions I heard; one was said to have chased down a four-wheeler and ripped the fender off with its teeth as if biting into the haunch of a fleeing reindeer. These sounded like monster stories to me, what they tell kids and visitors like me to keep us from wandering out too far and getting into trouble.

That particular day, the village council gave me permission to go more than three miles. Tonight I bumped it up to four. Drawn into the swirl of clouds and green earth, forgetting the sea behind me, and the village, I kept going.

Island residents are mostly Siberian Yup'ik, a subsistence culture that has lived here for at least two thousand years. St. Lawrence Island has two villages, Savoonga being the nearest, population 650 or so. I found that people shot, trapped, or threw nets at almost everything that moved. A Yup'ik hunter told me that the wolves were particularly difficult, retreating into the island's interior to avoid being trapped or shot. Polar bears? They knew what to do with polar bears. A polar bear came too close to the village a few weeks earlier and was now a robe hanging behind one of the houses, grease turned to amber, eyes like buttonholes. Wolves were harder to kill.

Grizzlies and polar bears lived this far north in the Ice Age, but the big inland bear was the now-extinct *Arctodus simus*, the short-faced bear. Its forelimbs were elongated, with a grip large enough to wrap around a muskox, or horse, or a juvenile mammoth. The New World was its home turf, a new animal for people moving west to east.

I felt that I'd been here before, had walked into these grassy slopes on a sunny day, horses in the distance lifting their heads, watching me pass. Wildflowers would have been blowing in a warm breeze. I would not have sauntered along like a nimrod, though, hands in my pockets. I would have been more alert. Early Siberians inserted stone blades into the shafts of their spears, increasing the depth and ease of penetration into walls of flesh. No chewed-on human bones have been found up here, though I can't imagine much would have been left after an attack from a four-hundred-pound scimitar cat or a pack of Beringian wolves.

Something moved far ahead. I stopped and squinted, tried to pick it out, but my eye wasn't accustomed to these green waves of tundra. Near and far looked the same. The storm obscured any sense of distance or scale. It was animal, that much I could tell. It had a *canid* appearance, dog-like, tracking back and forth. All the dogs were chained up back in the village, waiting for winter. They wouldn't be out here. Was it a fox? It was hard to tell from this far away.

I glanced back at the village. The roofs of Savoonga huddled at least four miles behind me, like urchins at the edge of the sea. One-story houses gathered around a school with a gymnasium, and no road access to the rest of the world. Whale-hunters. Seal-eaters.

I walked on, losing the movement I glimpsed, picking it up again. The animal split into two. My eye struggled to smooth the distance. Wolves, I thought. I was seeing two wolves.

I had thought the people in Savoonga were kidding about wolves. *Canis lupus* was extirpated from this island around the 1920s. When I asked an old man in front of a woodstove about this new appearance, he shrugged and said they must have come over from Siberia, crossing on sea ice the last time it froze solid enough, a decade or two ago. I looked at the animals' tails: Foxes move with their tails down, dusting the tundra, wolves with their tails raised. These tails were raised. I suddenly understood what the people in the village had been talking about. I was unarmed and had no idea how long these two animals had been heading in my direction.

In Savoonga that day I'd been given a small, colorful slip of glossy paper that read "Land Crossing Permit." A representative from the council ran into me in the street, pulled it out of her jacket pocket, and told me the council had decided I was worth the risk. I'd been hanging around the village long enough that they must have decided that I was OK, or that at least if I were attacked by animals, it wouldn't be from a lack of warning. I was researching a climate change project, using the island as a base for writing and interviews, taking in local anecdotes describing new species in the waters, stormier summers and winters distressingly warmer. I was told of coastal landmarks and beaches going under year by year. I slipped the permit into the space in my wallet where a driver's license is supposed to be. I now had permission to walk across the Bering land bridge, given by the oldest people on the land. Honored, I took it with a small nervous gulp that I hoped the woman hadn't noticed. If I were ever asked if I belonged here, anywhere in the Americas, I could open my wallet and say, *Why, yes, I do.*

I kept walking, figuring I was better off if the wolves didn't notice me noticing them. If they were indeed wolves, and not a pair of half-feral sled dogs escaped and running back to the village—animals I'd rather not encounter either. Wolves might be curious, never allowed to get this close to a human. They could see I had no rifle over my shoulder. They understood me by the way I walked.

What did I have to defend myself? Fingernails and teeth? A loud voice? My mom's half-sandwich? Dread sank through my stomach and settled into the notch of my groin. Out here, I was naked, no bush or boulder to duck behind.

In the Ice Age I'd be adrift in blueberries and wildflowers. Looking up and seeing a pair of Beringian wolves loping toward me, I may have had a similar sinking feeling. Beringian wolves were boxier than modern wolves, larger skulls, necks a few inches longer, heavier muscle connections, more robust teeth to handle the larger prey of the late Pleistocene. I would have been prepared, at least had a knife or blade, a sharpened, bifacial stone wrapped in leather, a dog, a buddy, a tribe. I would not have been alone like this, four miles from anyone. This would not have happened.

They looked leaner than the wolves I'd been told about. Blond, patchy brown, tails lifted, they lived in the rugged middle of the island. I let myself down into a gentle draw and as soon as I was out of their sight, I spun and started swiftly back for the village. Foxes, dogs, or wolves, it didn't matter, I made tracks out of there.

The kind people of Savoonga had offered my mom and me a place to stay, a pressboard second-story meeting room where my sleeping bag was already laid out on a cot. She was probably sighing in her sleep. I thought ahead to the schoolroom clock on the wall, the microwave sitting obediently next to a stack of cups, and my socks laid out to dry on a piece of roller luggage. I wanted to be there instead of here. I wanted what everyone wants, to be protected from the trials and fears of the unknown. Most of the world's large predators have been exterminated or driven to the far corners. We have all but forgotten how to inhabit this kind of fear. We gave up spears and skins and the weather on us day and night for cup holders and cell phones and doors to close behind us. What, I wonder, was lost?

A mammoth kill on the Russian Plain left part of a spearpoint in a rib. Judging by its embedded depth, considering the skin and muscle the projectile had to penetrate to reach bone, the spear must have been thrown from about fifteen feet away, proof that people got very close to excessively large animals.

My mouth felt dry and I thought, *My god I've grown soft.* I could scarcely imagine myself approaching any animal with no more than a sharpened stick or a stone point. That kind of involvement with wildlife has escaped most of us. Now, at the slightest sign of canids, I was hightailing it for home.

When my cover played out, I stopped and looked back at the fog-obscured terrain. Distances turned in and out of each other, and I saw the wolves again, closer. I fell into an easy pace, not wanting to be seen in a spotting scope from the village, a fool running with his hat clamped down on his head, fleeing from a pair of curious wolves. I knew wolves mostly from the Lower 48, night howlers in Wyoming and Montana; I still listened for them in the backcountry of Arizona. But those wolves were hemmed in by a human tapestry of roads and towns. What did I know of hungry Siberian wolves stranded on an Arctic island? I continued through draws and hills until the village was a couple of miles away. My view had planed out, elevation flattening, houses on top of each other behind the gravel slash of a landing strip, shipping containers stacked along one side, towering white wind turbines. The wolves, if I were brave, or the foxes, if I'd been duped, had turned away. I felt remiss for how little I knew about wildlife on the tundra, how much animal intelligence I've lost. At least they knew better than to come any closer. There'd be four-wheelers and rifles in no time.

I teetered for a moment on that tightrope. I wanted to walk back into the open and see whether the two animals would pause and look back. I wanted to feel the tang of their observation, to remember what it's like to be human.

Arriving in the windy gloom of Savoonga around one in the morning, still light out, I passed kids on their bikes. A man came from a plywood house to show me ivory harpoons dug out of the ruins of ancient, abandoned villages. Another man coming from another direction had pieces of mammoth ivory that were found in the tundra around here.

At town hall, bingo was just letting out, people mounting their four-wheelers for home. In one of the only two-story buildings in the village, I turned up a flight of worn wooden stairs and opened a door onto the quiet room where my mom and I were staying. Microwave. Styrofoam cups. Coffee pot. My mom was asleep. I sat on my cot to unlace my boots, and she stirred. She cracked an eye and asked if I had a good walk. I said I did. I did not tell her how far I still had to go.

The first people to enter Tasmania in the South Pacific thirty-six thousand years ago—becoming the southernmost Ice Age hunters—ate wombats and wallabies and used bone points hafted onto wooden shafts. In Borneo about the same time, Paleolithic hunters fashioned stingray spines into barbed spearpoints and used them to take down wild pigs, monkeys, turtles, and civet cats. Farther north, people hunted larger animals, and their tools became more robust, made of knapped bone and rock to match the size of their prey. At an Upper Paleolithic site in southern Portugal, fire-cracked rock and stone anvils have been found among the systematically smashed bones of deer, bison, reindeer, and horse, evidence of marrow rendering. The North was a harder place to survive, requiring adaptation to deep winters.

In the Ach Valley of southern Germany, young, nursing-age mammoths were hunted in the spring and early summer. A cave there produced the remains of a cave bear killed by human hunters. The beast would have been fifteen feet tall on its hind legs. The tip of a flint spearpoint was found stuck in its vertebrae just under the shoulder blade. The blow would have come from its right flank, probably aimed at the heart or lungs, but stopped short. It was not a killing blow. More spears and spear-thrusters must have been on hand. The bear's skeleton was found scribed with butchering marks, the robe of its fur taken off, signs that humans had won the day.

People were well established in parts of Siberia by thirty thousand years ago. Crossing the land bridge, even if they had no idea what lay ahead, they would have known that the country was changing—the first appearance of short-faced bears, the last sight of woolly rhinos. Even the beetles on one side were different from the beetles on the other. People would have known they were entering a new world.

When my mom and I flew back to the mainland, we landed in Nome, Alaska, where the land bridge once ramped onto the upper left-hand corner of North America. Here, people would have seen the first topography since the peaks of St. Lawrence Island, the land rising ahead of them into bright glaciers and rock outcrops. The treeless mountains standing above the town are visible fifty miles out to sea, out to land.

A broken length of woolly mammoth tusk was found about a hundred miles northeast of here near the coastal town of Kotzebue, Alaska. The tusk, which dates to thirty-six thousand years ago, had been scored by stone tools. One archaeologist described it as a "cutting board." Its antiquity may be irrelevant; the tusk could have been found and cut much later. This is where dates become hard to track. What is a sign of people, and how old might it be? Not far west of this find, among the tors of Serpentine Valley on Alaska's Seward Peninsula, fluted spearpoints for hunting megafauna have been found, along with charcoal from willow fires with dependable radiocarbon dates around twelve thousand years old. Beyond that, in a cave in the Yukon Territory, a butchered horse jaw has been dated to twenty-four thousand years ago, the bone cut when it was still fresh.

The land bridge remains a hypothesis. Though early people are found on both sides, no physical artifacts or sites have been discovered to prove that they crossed through here. Atlantis has sunk and taken the artifacts of a thinly spread Ice Age people down with it. Let's say, for argument's sake, that the first people walked across the land bridge from Siberia and set up camps and seasonal territories on its expanse for many generations. For anyone heading east toward the rising sun into more of the same grassland steppe and winding rivers, the mountains above Nome may have been the first glimpse of North America.

Just above town stands Anvil Mountain. Four massive radio antennas have been placed there, defunct remains of a Cold War communication network, good views of the Bering Sea as well as inland. This is the White Alice site, a former sentinel for aircraft and ballistic missiles coming through polar airspace. These aren't spindly antennas with red lights flashing on their tips; they are large, corrugated parabolas like drive-in movie screens.

One pair now faces the sea, with Russia just over the horizon. The other aims into the rest of Alaska, Canada, and all of North America, a tropospheric bounce from here into the rest of the United States. Signals of incoming weaponry went to Southeast Alaska and jumped to an underwater cable that could get a message down the West Coast and across to Washington, DC, in half an hour, following a potential migration route of the first people, a convenient corridor of travel along the coast. Another signal was sent through the Canadian interior, also an ancient migration route, as if the land keeps offering a way and we keep taking it.

Shut down in 1979, replaced by satellite telemetry watching for nuclear launches, this site no longer communicates. Concrete bases are cracked and invaded by tundra. Bolts holding down the metal struts of each parabolic superstructure are dull gray from oxidization. The place was empty but for wind and a distant herd of muskox grazing among willows, their winter fur raked off by branches. The summer sky was clear around us, the air crystalline, light within light, as if we were inside a bright Arctic seashell.

My mom waded into the gnarled, thigh-deep willows. She filled her pockets with soft muskox wool, called *qiviut*. The act was so old, so natural, I don't think she was aware of its significance. The wool felt good in her hands, warm on her cheeks. She couldn't help reaching out for it, as if the impulse ran in her bones, in her ancestry, from the time we used to do this, not just Beringians but all people. It wasn't that long ago that this was how we spent most of our time, fingers plucking wool from bushes, pecking one stone with another, watching the sky for countless lifetimes, the sun circling with no end.

In the late Pleistocene, my mother would have carried a basket or woven sack, filling it with coarse, reddish mammoth wool, or maybe the pale wool of Pleistocene horses combed off by twigs as they browsed the willows. Everything up here was shaggy. Scimitar cats and American lions, the great short-faced bear *Arctodus*, all wore heavy capes for winter and would have been mangy in the warm Ice Age summers, the ground clumped with their shed hair. My mom would have worn skins and leggings and carried a pouch of small stone blades of Asian design. Eyed sewing needles made from mammoth ivory have been found on both sides of the land bridge in contexts from 24,000 to 14,500 years ago. She would have used the small blades to cut skins, and the needles to sew tailored, tight-fitting clothes, the things that had to be invented as people moved north. If she were anything like the first robust signs of people in the Siberian Arctic, just across the land bridge twenty-four thousand years ago, my mom might have been adorned with bone beads or the teeth of animals drilled and strung on a cord. These were people with preferences and sensibilities. Curtis Marean, at the School of Human Evolution and Social Change at Arizona State University, told me, "A child from twenty thousand years ago would be a fully modern human, well within the range of variation of modern humans today." That means they would have looked like us and been able to act like us. An Ice Age kid might have navigated public school as well as any. If we saw one of them strolling around the White Alice site—a Paleolithic daydreamer dressed in a jacket and jeans—and he or she waved to us, we would wave right back. These people came here with languages and customs, body adornments, styles of weaponry, tailored clothing, pole-and-hide structures, and burial rites involving red ochre. They were not a blank slate.

Diane Hanson, the head of the Anthropology Department at the University of Alaska in Anchorage, warned me not to assume that this means we're the same as them, however. She and her colleagues make the same mistake. "Have you read our papers?" she asked, telling me that the first people are written about from a dry, academic vantage. "We tend to make them look like us, like an academic department, a bunch of people in their thirties sitting at their desks waiting to publish."

As my mom browsed through the willows, scooping up muskox wool, I climbed on top of wind-crazed outcrops of schist, looking down to the blue arc of the Bering Sea. It felt as if the great shell of the sky were opening.

I wandered back through willows around the base of the towers, ducking along muskox trails in the warm hairy scent of beasts. A muskox was still here, its head down grazing, teeth snapping at summer grass. It lifted its gaze, and two boxy brown pupils stared at me from so close I could have tapped the animal with a stick. I'd never been this close to such an unfamiliar animal, all head, haunch, and horn. I figured I'd hear one of them first, not that I'd walk smack into it. I'd encountered a muskox in Greenland, out with a couple scientists at the edge of the glaciers. We scared up a bull on a ridgetop. It huffed and stomped about fifty feet away, hooves chopping the snow, an encounter that gave an entirely different impression than the one beneath the radio towers. This was at the end of a dirt road out of Nome, the wildness diminished. Grass hung from its mouth, and it blinked.

The animal did not care that I was here. It let out a huff through long, flattened nostrils, and lowered its head to eat.

The muskox knew my kind, accustomed to the dirt road to White Alice, and the occasional harmless, useless visitor. I was part of its daily machinations, hardly worth thinking about.

The beast's eyes were slotted and darkly marbled, the mustache above its nostrils damp from its breath. Its head was like a great anvil, its body more haunch and horn than back or hind legs. I stepped backwards, and backwards again, retreating down the trail. The sharpness of my vision in that moment may explain why Paleolithic caves of Spain and southern France have been painted and scratched with animal shapes rendered with an artist's accuracy and flair. Those encounters were cleaner than what we have today. Everything about the animal would have been located instantly—angle of horn, flatness of mane, legs at a run, or body standing still. The presence of giant animals was life and death, and they were everywhere. In the cave paintings, the horn of the woolly rhino was given echoing outlines that convey movement, drawing your eye, saying, *This is the part you have to watch out for.* The way we now speak in texts, or remember the turns on and off ramps during morning traffic, that is how they once grasped animals. There was seniority on the land, and it was not human.

2

INNER BERINGIA

25,000 YEARS AGO

In my twenties, I left my job reporting for a weekly newspaper in a small Colorado mountain town. My career as a journalist lasted two years, long enough for me to save up and hit the road for Alaska with a college buddy. All I left behind was a post office box and the tipi in the woods where I'd been living.

Todd Robertson and I had guided summers for a dubious wilderness canoe outfit around Wyoming, Colorado, and Utah. We knew something about rivers and had a canoe tied to the top of his truck. That was our fortune. We were free to disappear.

It was 1992. In the town of Whitehorse, Yukon Territory, Todd and I launched our blue canoe and fifty days' worth of gear into a Yukon River freshly purged of spring ice, as far out on the planet as either of us had ever been.

I think about mammoth skeletons found in the bottom of a South Dakota pit, where more than sixty juvenile and adolescent male Columbian mammoths stumbled at different times, becoming mired in warm spring mud they could not escape. But no female mammoths were found in the bone assemblies. If mammoth social structure resembled that of modern elephants, these young males did not belong to a herd with a dominant bull, moving with females and their offspring. They were young, energetic outcasts looking for a home. All those skulls and tusks jumbled around at the bottom of the pit—I imagine the adolescent males trotting along one at a time, tusks lifted proudly in the air, before they tumbled over the edge and out of sight. Todd and I were like young mammoths, itching to go.

Not that wandering great distances is solely a male occupation. About five hundred miles downstream on our journey we encountered an all-female raft, dreadlocks and tans, the butt of a rifle sticking up from their gear. The need to get the hell out there is a shared trait, but young males, based on the South Dakota mammoth site, seem to be more likely to end up with their bones jumbled at the bottom of a hole.

Several hundred river miles below Whitehorse, the Yukon left the mountains and spilled across the Arctic Circle. We were bug-bitten and bearded, melded into our canoe as if we lived in it. Inner Beringia opened wide, nothing but sky, river, and scrawny island forests ahead. Almost a month of mosquitoes, bears, and wild water through canyons lay behind us, where inundated forests had been collapsing into the river from spring runoff, their mast-sized trunks rolling in the current.

We were in a sprawl of channels and sloughs, the Yukon River poured across the flat Alaskan interior like a jug of water dumped on a tabletop. Six miles wide, the river was a labyrinth of braids and spindled black spruce on islands with long, curving shores. The Arctic sun made circles around our heads like a tilted hula hoop. It was hard to tell just by looking if it were nine at night or five in the morning, sunsets and sunrises drawn out for hours, inseparable.

In the flats I insisted on taking the front of the canoe. From the bow, I didn't have to look at our gear strapped into the boat, or the back of Todd's baseball cap, a thing I'd been staring at for weeks. I wanted to be first, the prow-head of a boat cutting into the unknown, the sky inhaling us.

But we were nowhere near first. Before us came Athabascans and Inuit, and the linguistic roots of Uto-Aztecan people. There were island-hoppers and land-bridgers from Asia. Mammoth-hunting camps cropped up on the Tanana, the next big river south of the Yukon, while up the Yukon's tributaries, human signs have been found dating back to at least twenty-four thousand years ago, some say forty thousand years. At the time I didn't know about these older dates or sites. The word *Paleolithic* was not on my tongue. I knew we weren't first, but it sure felt like it.

Holding onto time on the river was difficult. I'd pull out my pocket watch and look at its hands as if they spelled out an arcane message. The lack of mechanical time made space seem transparent. There was no Alaska, no North America. Earth wasn't an orb in space. It was a turtle swimming through a primordial sea, its shell so old and enormous it had grown mountains and glaciers like barnacles.

The arrow of early American migration hooked through the heart of *Mammuthus primigenius* habitat, country of the woolly mammoth. This animal's range was defined by steppe, which stretched from Paris to New York, Asia and Beringia in the middle. When humans began to appear on this side of the world, the Laurentide Ice Sheet was for the most part welded into its mountainous neighbor to the west, the Cordilleran Ice Sheet, which blocked the way. During warmer parts of the Ice Age, when ice sheets separated, a corridor opened between them and allowed the woolly mammoth to pass through the cool grasslands and exploded glacial kettles of Alberta, into Montana and the Dakotas, opening onto the rest of the Americas.

Homo sapiens and *Mammuthus primigenius* occupied the same

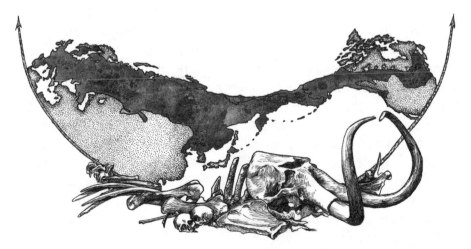

Woolly mammoth range, darkly shaded: a continuous grassland ecosystem from Portugal to New York, which also encompasses the range of Beringians who utilized the same habitat getting from Siberia into North America

ecosystem, sharing a jigsaw puzzle of summer grasses, skinny trees, distant ice caps. It was one long country and it stretched from Western Europe to the American heartland, from France and Scandinavia to the Great Lakes region and the Northeastern Seaboard of North America, bound in the middle by the land bridge. Tusks and skeletons are regularly found in the permafrost of the Far North, where Siberian fossil hunters use pickaxes and hoses to blast them out of cutbanks along rivers where the animals once grazed and died. These are the same places where some of the first humans appeared, following the mammoth, or at least following the biotically rich steppe where the great beasts trod.

The body of a mammoth calf with strawberry blond fur was found preserved in Siberian permafrost, while the adults appear to have been more coppery and brown. Mammoth fur consists of two distinct layers, one composed of long, coarse guard hairs, six times thicker than human hair and up to three feet long, and an inner layer of shorter, softer, thinner hairs for insulation. Too heavy and dirty for most uses, it's not a hide you'd necessarily cut into pieces and wear. There would have been plenty of other animals to use for hides: caribou, horse, deer, bison. Muskox would have been hunted and caped, their fur lighter and better adapted to cold than mammoths.

A major mammoth excavation in Siberia is the Berelekh site, near the mouth of the Indigirka River on the Arctic Ocean coast. This was a dying ground, a Siberian graveyard of bones including the remains of at least one hundred and twenty individual mammoths. During lean times, starving elephants are known to gather in places where food or water is more easily found. Sometimes they die there. This appears to have been the case for the woolly mammoths at this Siberian site. Excavators used a water pump to melt permafrost and found reddish-gold wool and mats of guard hairs. They exposed partially rotted flesh and a well-preserved mammoth hind leg still bearing skin. Ninety-nine percent of the remains they found came from mammoths, the rest from cave lions, woolly rhinoceroses, horses, reindeer, bison, and more than eight thousand willow grouse, estimated by the plethora of tiny bones.

No human kills have been found at Berelekh. The mammoths appear to have died on their own. The humans came here like any other carnivore, drawn by the smell of meat, scavengers like the bears, big cats, and wolves. They left blade-like rock flakes and thin stone knives, and some of the bones of the mammoths were cut where meat was taken off. Among rotting rib cages along the banks of the Berelekh, people left behind a handful of small pendants, polished stone with drill-holes for wearing or hanging. An artifact hunter found a piece of mammoth tusk upon which a ghostly, long-legged mammoth had been etched: an early artist scratching on a chunk of mammoth bone. I see her or him in an autumn chill, grinding into the bone as willow grouse clucked and rattled their throats in evening light.

Artifacts at Berelekh appear to have been left around twelve thousand years ago, a later occurrence than the very first people. The teardrop-shaped stone projectiles found at the site are more Alaskan in origin than Asian, suggesting the door of the land bridge swung both ways. It was all one land, one ecosystem, no reason to go east any more than west.

Sergey Zimov, a Siberian ecologist working to preserve the remains of Mammoth Steppe in Russia by introducing natural grazers and removing humans, wrote in the journal *Science*, "In the mammoth ecosystem, the collective behavior of millions of competitive herbivores maintained the grasslands. In the winter, the animals ate the grasses that grew the previous summer. All the while they fueled plant productivity by fertilizing the soil with their manure, and they trampled down moss and shrubs, preventing these plants from gaining a foothold."

Zimov believes that if mammoths could be brought back through genetic miracles to repopulate the Far North, they could help restore one of the world's largest relic ecosystems.

At its prime, populated by mammoths, sloths, and bison, the Mammoth Steppe was a rapid processor of CO_2, which helped cool the Northern Hemisphere. Among big herbivores, one new calf a year can increase the population a hundred times in twenty to twenty-five years, enough to knock down trees and mow grasses to the ground. That causes increased reproduction among grasses, a feedback loop leading to a global pasture. This rich, biologically productive, planet-cooling biome took up most of the high Northern Hemisphere, working the atmosphere like a machine.

All of this on a turtle's back.

Native American DNA split from its Eurasian roots around thirty thousand years ago, while in the New World the oldest genomes to have been discovered come from thirteen thousand years ago. Between these dates, the first people seem to have been floating out there, little to no linguistic or genetic contact with the rest of the world. This has given rise to the Beringian Standstill Hypothesis, the idea that the first people simply stopped up here before they entered the Western Hemisphere, balanced at the top of the world: Siberia, Alaska, northwest Canada. Adapted to cold conditions at the northern heart of the Mammoth Steppe, they may have waited to go forward into the rest of the Americas for thousands of years, in which time people learned of mammoth graveyards and used different technologies telling one age or region from the next.

Pollen counts from the time reveal an abundance of wildflowers, grasses, and island ecosystems of birch and spruce, all of which were favored by Beringia's mammoths and giant sloths. For the most part, the path any farther into North America was blocked by ice. During warmer ages, corridors opened between ice caps, letting bison and mammoths pass, mixing the animal genes of two hemispheres, like taking big, deep breaths between sides of the planet. Of course, humans were caught up in one of these breaths.

When early humans appeared in Alaska and neighboring Yukon Territory, between about twenty-four thousand and fifteen thousand years ago, ice was at its maximum. Humans may have been drawn east out of Siberia, but when they got here, the door into the New World was closed. The far side of Beringia, the Alaskan interior, would have formed a glacial wall with tongues of ice lapping across wind-blasted tundra and bare rock. You'd be better off staying with the rivers and giant animals of Inner Beringia than venturing into this white oblivion, nearly two thousand miles of ice to the nearest landfall and, during the last glacial maximum, no easily navigable coastline.

The question shouldn't be, did the standstill happen, but rather, why didn't they stay here longer? You could get lost out here, so much country, so many rivers winding into each other that groups could have gone years or entire generations without seeing others. There must have been lost clans and hermits, people up in caves forgetting the rest of the world. After eight thousand years of living this way, if the Standstill Hypothesis is correct, new linguistic and genetic entities would have formed, Beringia becoming a womb for a new people.

I did not know it had been a womb. To me, it was a river with a sky stretched over it. If it was giving birth to anything, it was more horizons. We slept when the sun tipped near the ground, our two tents set on gravel bars for the long sunset. This low sun gave us only a few hours until it baked us out again and sent us back to the river. For twenty hours at a time, we paddled, slept, pissed into cans, and watched horizons rise ahead and fall behind. The only other people in our world were animals. We became familiar with the sound of snorting moose as they swam wide channels for other islands, the crowns of caribou antlers rising through the tops of willow shrubs, a grizzly standing on a faraway flood-stripped island on all fours, watching us with mild interest. We were passing through territories with boundaries we couldn't see, herds and predators, the long exploratory howls of wolves rising from the taiga.

The first known human architecture appears in Siberia by twenty-three thousand years ago, when the only way to survive year-round without having to migrate south was a well-built shelter. Circular huts were constructed, twenty-some feet across, fire hearth near the center, hides draped over wooden posts and beams, their bottoms held down by a ring of caribou antlers or mammoth bones. Todd and I carried tents framed with plastic bones and thin metal poles, bug screens and nylon. To Pleistocene travelers accustomed to hides on wood frames and floors dug into the earth, peeled down to permafrost, our tents would have looked as if we were draping ourselves nightly in gossamer. What would have been the use? People have different needs at different times. Our tents kept mosquitoes out. Theirs warded off giant bears and Beringian wolves, a piece of ground to defend, the difference between travelers and inhabitants.

Mammoth stories still exist in northwest Siberia, recorded by ethnographers interviewing Ugric and Samoyedic speakers in the early nineteenth century. The beasts are known as *maa-xar* or *muw-xar*, translated to "earth bull" and "earth horn." Shamans are said to have ridden them into the underworld, and in some stories they were forced into underground exile. Mammoths were terrible and destructive creatures, they ripped up forests and laid waste to mountains.

When they come to the surface they die, turning to bones. This is why no living mammoths are found.

An anthropological transcript taken from the Ugric tongue in northwest Siberia reads, "Nowadays we do not see mammoths any more. The Sky God has doomed them to death. Only their horns, marrow, bone and ribs can be found in sandy river banks."

Too ruinous for this world, they were driven underground. I imagine them now huddled in the dark beneath our feet, making hardly a sound as they listen patiently to the ground above, waiting.

They would have heard the quickening thaw of permafrost and the hiss of rivers carving through banks, releasing their buried tusks and massive long bones. The size of animals has decreased significantly since then, proboscideans gone from the North. Our bears look like pets compared to long-legged *Arctodus*. The world has changed.

As the Yukon came down from its spring and summer runoff, cobble islands and bars revealed themselves from under flood stage. Uprooted trees grounded and gathered into logjams at the upstream ends.

One evening as the sun ran low, we landed on a naked, teardrop-shaped island surrounded by a quarter mile of swift, cold, muddy water. We figured we didn't have to worry about bears, so we could put away our dainty bear bells. For most of the trip, we'd land on high alert, never cooking near where we slept, switching out dinner clothes for sleeping clothes, hanging our food in trees if we could. Forced into the woods by spring runoff, we'd been sleeping in the nests and trails of large bears. But islands like this seemed safe. As we started unpacking and putting up our shelters, a large black bear, *Ursus americanus*, ambled up a far shoreline. We watched it, then went back to setting camp, glad to be across the current from the bear. Maybe three hundred pounds, it studied the ground at the river's edge as if it had lost something. Its movements seemed anxious, preoccupied.

We'd had good relationships with bears so far. For that we were grateful, especially since we carried no firearms. Todd hated guns, and though I grew up around my dad and weapons and knew how to aim and shoot, I thought it was somehow more noble, closer to nature, to face animals on their own terms.

The bear jumped off its cutbank into the water. The loud crash reached us on the island, and we looked up from camp. The bear began swimming a ferry angle, positioning its nose upstream into the current. The calculation didn't require much thought: current and distance divided by the power of the swimming stroke. The bear was angling for us.

In the Lower 48, we'd scare away black bears by banging pots and yelling. This animal wouldn't give a crap about that kind of racket. From across the channel, it could smell that we were unarmed; if it ate us, there were no other humans nearby to seek retribution. Where we were camped, how we unloaded gear, the way we stood staring blankly as it swam toward us—we told the bear exactly what it needed to know.

First priority in Paleolithic minds: Don't get eaten. Opportunities for a messy demise would have been everywhere. This is part of our shared human history. Our jaw morphology, musculature, and omnivorous teeth are not like those of, say, cats or dogs. We are more utilitarian, part predator, part prey. We ate plants and insects, but we also knew how to fight and kill. In their book *Man the Hunted*, anthropology professors Donna Hart and Robert Sussman describe our human ancestors as "just one of many, many species that had to be careful, had to depend on other group members, had to communicate danger, and had to come to terms with being merely one cog in the complex cycle of life."

Not much has changed in our makeup. Only the pieces are rearranged. In your bones, you remember fight or flight: the possibility of claws holding you down, and the rip of jaws and teeth rending flesh.

Four choices:

1. Shout and wave, pelt the bear with rocks, cookware, or bean cans.
2. Jump in our canoe and paddle swiftly away, come back to face the carnage of our campsite later. That is, if we could battle our way upstream to reach the remains of our camp in the big, fast river.
3. Scramble to the far end of our island in the jackstraw of washed-up trees, branches sheared off by tumbling, and watch our fate unfold from there.
4. Do nothing at all.

The last response seemed best, the only thing requiring momentary paralysis. The bear was now fighting harder with the current, glancing over its shoulder at us, nose sticking up, paws paddling like big wet mitts. Every stroke was a grunt, working for extra purchase. We could hear the size of its lungs.

When we saw its eyes, it was clear what the bear was thinking. As it looked over its shoulder past us to the downstream end of our island, gauging how much beach was left, it realized it had miscalculated. The bear put more punch into each stroke, trying to catch up, but the current was too fast. Barrel-chested, it took in a huff of air, its vigorous swim reduced to a light paddle. The game was over. The bear sailed past the tail of our island in muddy, swirling water. It didn't gain another shoreline for about half a mile, and it did not look back at us. It climbed onto a bank with its wet, shaggy hide, and ambled into a birch grove and the matrix of its endless summer day.

For Ice Age people, our black bear encounter might have served as entertainment. Black bears didn't live this far north at the time; they were too small and easy to eat. Their Ice Age bones have been found along coastal Southeast Alaska, but not here in the interior. They might have been regarded the way we look at wolverines: dangerous, toothy animals but small, something you'd hope to fend off with a big stick. A black bear playing cat-and-mouse with people, with bean cans, knives, and bear spray—that would have been entertaining, something viewers could clap and laugh to.

Arctodus would have been more swift and final. It would have entered the current like a tank. The Paleo audience would have cringed as the great bear charged; they'd have thought about this so many times, it would have been pure muscle memory. Estimates in body mass based on the size of major limb bones commonly show *Arctodus* weighing around a ton.

Some scientists have suggested that the short-faced bear was not a meat eater, though. Isotopic analysis of teeth from the European cave bear indicates that this famed and feared Old World ursine was almost strictly herbivorous. The same, they suggest, could be true of *Arctodus*, whose closest living relative is the spectacled bear of South America, primarily a plant eater. I visited Dr. Sue Ware, a forensic

Arctodus skull, the American short-faced bear, head
more cat-like than most modern bears

paleontologist specializing in carnivores, who scoffed at the idea that
Arctodus had a diet anything like its modern South American relative. She showed me an *Arctodus* skull that would take up most of the
passenger seat of a car. Ware opened the jaws, revealing a full set of
meat-slicing and bone-crushing teeth, somewhere between those of
a hyena and cat. "This was no herbivore," she said. She invited me
to try it on, and I opened the jaws and stuck my head inside. My own
skull fit perfectly. It could have clamped down on the top of my head
and whipped me against the ground like a fish.

If Todd and I had dived into the river to escape, *Arctodus* would
have been on us almost instantly. The one-ton bear would have bat-
ted us back to shore as if swatting salmon out of the water, then held
us down one at a time. We would have felt meat being stripped from
our bones. Or we would have felt nothing, our minds knowing when
it was time to go silent.

Morphologically speaking, the short-faced bear was probably neither pure herbivore nor pure carnivore. Specimens from Nebraska,
California, and Alaska have cranial and dental formations that com-
plicate the hypotheses. It appears that they were neither fast-running
super predators, plant grazers, nor dedicated scavengers, but rather
all of the above. Researchers comparing the traits of jaws, teeth, and
skulls called *Arctodus* "a colossal omnivorous bear whose diet probably varied according to resource availability."

Its actual diet has been revealed through bone chemistry. Isotopic composition of short-faced bear collagen tells us what the animal ate, and Alaskan short-faced bears ate mostly meat with some vegetation. They ingested the meat of not just camel, mammoth, and horse, but large carnivores, canids, and cats. The bear was both scavenger and hunter, subsisting off meat augmented by a salad of steppe grass, wildflowers, and berries. In many ways, it ate like a modern Alaskan interior grizzly. But it was significantly larger, a big, toothy welcome to the New World.

I felt the Porcupine River come in from the right, but I didn't see its confluence. The water changed color, became slightly clearer with less mountainous debris. The current swelled and picked up speed, given more mass and momentum, but the merger was invisible among big, open bends and islands. Rivers seemed to become each other. I'd heard its name and saw it on a map; I knew that somewhere nearby was another thread coming in from far away.

The Porcupine starts in northwest Yukon Territory, Canada, five hundred miles upstream of where we were. Its headwaters are a filigree of oxbows and winding channels among muskeg and spruce, a place called Old Crow Basin. On the side of the basin, a limestone outcrop stands over broad taiga, a high point looking across the circle of the Far North. This outcrop holds three small rockshelters known as Bluefish Caves, the site of what is possibly the oldest human activity on this side of the world, a likely place to end up if people were following the biggest rivers off the land bridge upstream into the continental interior. Thirty-six thousand mammal bones were recently analyzed from in and around two of the caves, and researchers reported that "wolves, lions and, to a lesser degree, foxes were the main agents of bone accumulation and modification, but . . . humans also contributed to the bone accumulations in both caves."

During the late Pleistocene, Old Crow Basin was a megafauna refuge. Big animals waited out the coldest, most brutal climate trends here, leaving permafrost cutbanks and cave floors littered with their bones. People would have smelled this at the meeting of the rivers. The Yukon would have been highly variable, flowing out of the folded arms of the ice sheets that capped half of North America. The Porcupine flowed out of lakes and muskeg, its headwaters curled up in the Beringian interior. It would have smelled alive compared to the ice-dominated Yukon, and the first people would have detected bones and grass in its waters, hints of mammoths and short-faced bears, graveyards of tusks, and herds of Pleistocene horses. They could have read this much and more at a river's confluence, where I was only aware that a new river had come in.

Old Crow Basin has produced flakes of stone tools, the oldest of them representing the Dyuktai Complex, from Dyuktai Cave in the Aldan River Valley of northeast Siberia. Sprawling river systems drain the Central Siberian Plateau, home of these Dyuktai artifacts. Hopping from one river to the next, people would have reached the land bridge and eventually the Yukon River, which would have lured them farther east, leading them up the larger Porcupine until they finally arrived in Old Crow Basin.

The Dyuktai tradition consisted of burins, scrapers, perforators, wedge-like cores, and pressure blades struck from the edges of stones. They first appeared in Asia around thirty-five thousand years ago. Jacques Cinq-Mars, a Canadian researcher of prehistoric lithic technology and, at the time, curator for what is now the Canadian Museum of History in Gatineau, Quebec, was the first to identify Dyuktai in North America. Cinq-Mars was the original investigator at Bluefish Caves in the 1970s and '80s, where he and his colleagues found stone tools and flakes mixed in with Ice Age fauna. They came up with small, sharp flint knives used to manufacture the tailored clothing and skin-and-hide shelters necessary for life in the North's cold climate. These microblades were like scalpels, ideal for cutting skins and hides into precise sizes and shapes. Cinq-Mars thought this looked like a Siberian technology.

He presented his discoveries to audiences that were unresponsive if not openly hostile. The going story was that the first people in North America were Clovis mammoth hunters who cruised in on the land bridge and filled the continent thirteen thousand years ago. Cinq-Mars had come up with a different story, finding human dates between twenty-seven thousand and twenty-four thousand years ago, not part of the popular model. His work was mostly dismissed. At conferences he was ignored, sometimes derided, his radiocarbon dates declared contaminated and unreliable. Old Crow Basin was not accepted as a location for the first people.

Bluefish Caves continued to produce stone flakes, hammered rock cores, and the remains of Ice Age megafauna, credible evidence that people were taking shelter here in the late Pleistocene. In 2017, four decades after Cinq-Mars began work at the caves, his pin went back on the map. A Pleistocene horse jaw from Bluefish, butchered by stone tools when it was fresh, came up with a radiocarbon date of twenty-four thousand years old. Fifteen Ice Age bone samples with clear human cutting and butchering marks, originally collected by Cinq-Mars, were reexamined by Lauriane Bourgeon, a doctoral candidate in anthropology at the Université de Montréal. Bourgeon subjected six specimens to radiocarbon testing and came up with dates between twelve thousand and twenty-four thousand years old, the oldest a series of long, intentional cuts on the inside of a Pleistocene horse's jawbone. These appear to be filleting marks, consistent with removing the horse's tongue, a likely procedure by hunter-scavengers twenty-four thousand years ago. Analysis of one of the teeth in this jaw indicates that the horse died in spring or summer, a warm season's meal.

Though this is a clear vindication for Cinq-Mars, it is still not the oldest date in the American North. Sign of people arises forty thousand years ago, a presence as faint as a shadow: a horizon of uniquely shattered mammoth bones in Bluefish Caves and the surrounding basin. The late Richard Morlan, an archaeologist who worked at the caves alongside Cinq-Mars and assembled the Canadian Archaeological Radiocarbon Database, found these bones and concluded they'd been shattered by early people. They must have used rocks, lifted them overhead and slammed them down on the bones, creating a distinctive break not seen in trampling or general decay. Such breaks are known as spiral fractures, and they form in bone only when it's "green," when an animal has recently died. No artifacts were found alongside the shattered bones, so most archaeologists have steered clear of them, and the early date has been given little credence. But the fact remains: Someone was probably here, a new agency or process at work on large vertebrates in the Old Crow Basin forty thousand years ago. Who they were, where they came from, and where they went remains a mystery.

Similarly shattered bones have been found in the American interior with dates old enough to say that people, at least hominids, appeared much earlier than is generally credited. The mixed DNA from half-human and half-Neanderthal bloodlines found in a forty-five-thousand-year-old human leg bone from Siberia suggests who these bone-breakers might have been: an early, primitive wave, a lost people of mixed race living as far away from the human core as they could get. They may have found the winding tributaries and lakes of the upper Porcupine River and traveled here seasonally to smash bones and scoop out marrow.

In a 1999 paper on their final analysis of Bluefish Caves, Cinq-Mars and Morlan wrote that "eastern Beringia offered no known ecological obstacle to colonization by people adapted to northern latitudes. Our interpretation of altered mammoth bones indicates that such a level of adaptation had been achieved by at least 40,000 years ago. We presume that people had spread across Beringia by that time."

Dyuktai technology is more than a shadow. Microblades and the chips of rock left over from their manufacture mark the appearance of the first identifiable technology in Canada and neighboring Alaska, what might be called the first fine-grained evidence of a people's presence. They were in place by twenty-four thousand years ago, not just breaking bones, but warmly dressed and braving winters in pole-and-skin huts, their tools and weapons refined as the earliest identifiable humans drifted across Beringia.

For me, there was no past. I was a minor god striding through the world, leaning from the bow of our canoe as if into a creation story. I could name things as I saw them, a colonist at heart. That is a bear, that is a cumulus cloud. This is rain, this is sunset. The world was opening to me, unknown and undiscovered as far as I could see.

On a sunny day, cumulus clouds rising far from each other, I was half asleep in the canoe, straw hat pulled over my eyes. Todd rested in the stern, paddle stowed, paperback shading his face.

I heard him sit up. He said, "Someone's out there."

I opened my eyes to a lattice of sunlight.

"Out on that island, waving," he said.

I put on my hat and pulled out binoculars. Across ripples of sunlight and water, my focus drew on a shirtless man standing on an island where he'd beached a canoe. I studied him for a moment. Black crop of hair, no beard. A small amount of gear, modest camp.

Braided Yukon River as it crosses the Arctic Circle,
six miles across from bank to bank

He looked skinny. He stopped waving as we took up our paddles and turned in his direction. He didn't appear to be Athabascan or Inupiat, did not look like a local. He had the gear of a long-distance traveler. When we were close enough, he waved excitedly with one hand, calling out *hello*. His accent was unmistakably Japanese.

We glided up to his island as he stepped into the water in laceless sneakers. He grabbed our bow, helped pull us to shore. I jumped out in river sandals. Todd, who'd put on a ratty T-shirt to be presentable, climbed over gear from the stern.

The man was weather-worn, face scabby, sunburned, about the same age as the two of us, mid-twenties, another young male mammoth trotting around the perimeters of the world. He said his name. I tried to repeat it, and he laughed. Todd tried next, and we smiled at our unfamiliar accents. I handed him my journal, and he wrote, MINURO SIGUNAMA in all caps. He said he'd been lost and we were the first people he'd seen in many days.

Minuro explained with his hands and broken English how he'd thought he'd fallen off the earth. One of the sloughs must have carried him out to the river's fringe, where he found himself cut off by a splintered logjam piled up from spring runoff. He climbed on top of it and looked across a blockade of locked and busted trees, filling the slough as far as he could see. He gave us an impression of a part of the river we hadn't seen, a dead end that would take days paddling against the current to escape. He was giving us a map of what he'd seen. We pointed back to where we'd come from, describing a sinuous array of channels, the river's inner mainstem, giving him our map in return.

He'd picked up a five-gallon-bucket of sticky salmon eggs back in Fort Yukon and appeared to have been subsisting on these alone. We'd already met another solo Japanese traveler before this, a man journeying by kayak. Forty-three days later we encountered a Japanese man who flew into Beaver from Fairbanks, a Buddhist coming to find the grave of another Japanese Buddhist, a famous man who married an Eskimo woman and lived out the rest of his days in the Alaskan interior. Now there was this man, alone in the flats. I wondered if Tokyo was so densely packed that young, adventuring Japanese were flying out like shooting stars, any reason to go.

We undid straps from our canoe and opened our stores, sending over half a brick of Velveeta, a processed food product that we kept at the bottom of the boxes. This was a substitute for a slab of fat we'd chop into our rice. We packaged up pilot biscuits and dried, salted seaweed for him, a half-gallon bag of jerky. He gave us salmon eggs, turning them with his hand like sticky beads, freshening his gift, which left Todd and me forcing our smiles.

A meeting like this in the Ice Age, spotting each other across the horizons of eastern Beringia, might have gone the same way. At the very tip of the original human expansion, people would have been few. As Lauriane Bourgeon's 2017 study of Bluefish Caves noted, "The archaeozoological evidence, together with the small size of the lithic assemblages, suggests that human occupation . . . was probably sporadic and brief." They may have come from entirely different geographies, Pacific Rim travelers encountering Beringians, coastal people who came up from the Japanese Archipelago meeting inland dwellers speaking a different language. They could have separated so long ago in the wilderness of Beringia that they'd forgotten the other people existed, noting only by tools or mannerisms that they either shared some ancient ancestry, or they did not. Conversation would have been carried out through gestures and food. If they were enemies, they would not have known.

Minuro tore off hunks of jerky from our supplies and ripped at dried seaweed with his teeth. His words and gestures explained how he'd come to this river once before with a canoe, and he'd been enchanted with horizon after horizon opening out before him. On his first trip to the Yukon, he'd gone back to Tokyo after a few hundred miles on the river. Now he was paddling the full length of the Yukon, another thousand miles from our meeting to the Bering Sea.

When we paddled away, Minuro stayed at his camp and kept waving. After half a mile, I could barely make him out. His island disappeared and the sudden isolation hit me like nothing I'd felt before. In our short time together, we'd become a small clan. This wasn't a loneliness I felt for myself, or Minuro, or Todd, but for the three of us together. We'd found each other in the Beringian interior and had again set ourselves adrift, the clan of people who fall off the edge of the world.

3

HOUSE OF ICE

An ice mass once covered half of North America. At its greatest extent, during the last glacial maximum between twenty-four thousand and eighteen thousand years ago, the solid glacier, miles deep in places, stretched from the Atlantic to the Pacific to the Arctic Ocean, with no way around it. About the size of Antarctica, its surface area was five million square miles. Anyone coming across the land bridge would have found the way ahead blocked. The permission I'd received in Savoonga to walk the land bridge was not *carte blanche*. It got me this far, but the path ahead was not so clear.

The view would have been of the Laurentide Ice Sheet breaking over the crest of the Alaska Range. The hulk of Denali, tallest mountain in North America at 20,310 feet, would have risen from a throne of glaciers, orders of magnitude larger than anything the people would have encountered anywhere in Beringia. The Alaska Range stabbing through an endless roof of ice would have lifted their spirits with awe even as it filled them with trepidation. Katabatic winds coming off the ice, generated by the intense cold, made the surrounding land almost uninhabitable, grasses replaced by tundra, rock, and thin, scrubby willow. The people would have entered a country gradually emptied of life.

If they climbed high enough onto the ice sheet, around the massive crevasses, they would have looked across an occasional summit sticking through the ice like a ragged bone, a mountain swallowed to its tip.

I once saw what they saw. I walked onto the edge of the Greenland Ice Sheet, looking ahead to seven hundred thousand square miles of ice and snow. I passed through sculptures of cornices and ice pillars stabbing upward. The last living thing I saw had been a muskox a few miles back. Ahead lay a smooth blankness with a few last mountains poking through. I wore an anorak and heavy boots, my face stung by chill wind. The sense of oblivion drew me; I could feel it bodily like a magnet. I'd live a day or maybe three without shelter out there. I imagined walking onto the ice, not knowing for sure if anything was on the other side. How could the first people have known that this wasn't the end of all creation? To some, it must have felt as if the world was finished and they had reached the final edge. How could they have known about the mountains and rivers beyond, the entire bulk of the Americas waiting for them?

Chances of survival on the ice could not have been reasonable when the next ice-free land was eighteen hundred miles away in eastern Montana and the Dakotas, nothing between here and there. That is, if you managed to stay on a direct line, navigating by stars as if you knew the way.

Eighteen hundred miles is how far Robert Falcon Scott's expedition had to travel in the austral summer of 1911–1912 to reach the pole and get back by sled and human power. Not one member of the final polar party survived, including Scott himself. The planks of their bodies, fingers knotted in the ice, were left in a tattered camp with dreary journals. One person had fallen into a crevasse and disappeared. Another, Captain Titus Oates, mortally weakened from frostbite, limped out of camp into a shearing whiteout, famously telling Scott before he did so, "I am just going outside and may be some time." His remains were never recovered.

With their then-modern tools and gear, Scott and his crew should have had an advantage over Paleolithic explorers, but apparently it was not enough. At the same time as Scott's ill-fated trip, the expedition of Roald Amundsen, a Norwegian explorer, reached the South Pole and returned without a single fatality. Scott relied heavily on the technology of the early twentieth century, while Amundsen followed the more traditional advice of the Far North's Inuits regarding navigation and equipment. Some ways of crossing ice are better than others.

When I told Richard VanderHoek, state archaeologist for Alaska, that I was heading to an icefield on the Kenai Peninsula to get a taste for crossing a glacial expanse, he asked whether I was doing it right. This was something that Paleolithic people had to have attempted from time to time, one side of their world dominated by ice. He wanted to know if I'd be carrying a shelter made of three caribou hides stitched together, or the hides of two female bison split with a stone blade and overlapped onto a willow frame. I told him, "No."

He said, "Caves are fine, they can be pretty damn limited. Rock overhangs are good if you could put a fire up against them."

I explained that I was going with a group and we'd be taking tents and sleeping bags. I was looking for ways and routes the first people might have used, but I was trying not to be a blundering fool about it. There would be no caves. We were crossing a segment of a several-hundred-square-mile icefield in south-central Alaska. Our fires would be liquid fuel stoves, looking like miniature lunar landers, not the bone fires of central Alaska during the Ice Age. Ridding ourselves of modern effects would offer no deeper insights into the Paleolithic, I reasoned. I'd probably die within a week without them. It would be like putting an Ice Age hunter behind a desk, eyes struck with confusion over the keyboard, the stapler.

VanderHoek shrugged and said, "Bison hides are lighter to carry if they're female, in case you change your mind."

"Is it safe for idiots?" I asked.

"None of this is safe for idiots," said David Stevenson, my friend and mentor, a soul-brother who would turn sixty while we were out on the ice. He said this as we looked across an empty expanse of mountains and glaciers, nothing but rock and whiteness before us.

A mountaineering author and book review editor for the *American Alpine Journal*, Stevenson was the de facto leader of our group of five heading for the Harding Icefield, one of the largest remaining ice masses in North America. He and I had taken a few quick trips and mountain climbs in the Chugach Range outside his home in Anchorage, finding good company with each other, enough to plan this venture. Stevenson was here to ascend his first *nunatak*, an Icelandic word for a mountain summit stabbed through ice, a castle of rock completely isolated. It was on his bucket list, and the Harding has plenty.

Stevenson and I assembled a small ice-crossing tribe, carrying rope, sleds, skis, and gear on the first load a few thousand feet into the mountains at the edge of the Harding, a giant glacier poured like latex into the crown of the Kenai Mountains. The neck of a ukulele poked out of the top of my pack, gift of a woman we met at the trailhead on the freshly snowed valley floor. None of us knew how to play the instrument, but we brought it anyway. With a pack weight of around eighty pounds each, what's the difference? Going into the unknown, talismans might come in handy.

It was late May. Forty-seven inches of snow had fallen across the Kenai the week before, and ahead lay frozen, treeless earth. I kicked in step after step, counting my breaths under my load, thinking like a Sherpa, or trying to: Fifty steps and rest, fifty steps and rest.

I stopped and leaned my weight over my poles, looking up at the two people climbing above me, their shadows flashing against a golden ball of sun, Stevenson breaking a steep, knee-deep trail for the rest of us. I looked down past the other two kicking their way up behind me. The ladder of our tracks disappeared below them, slipping over an edge. Two thousand feet below, a forested valley was melting out of the snow. A glacial river threw question marks across a plain of gray and rocky outwash. I could still make out the road we'd come in on, a meager ribbon to a dead end, a farewell.

People once traveled at the speed of two feet, or possibly a paddled boat, but no faster. Most of our evolution was at this pace, one step and the next. The way we zoom around in cars and planes has changed the way we understand the dimensions of geography. Distances are abstract, unapproachable; what you see when you open the shade on an oval window in an airplane and look down thirty thousand feet is walkable land, mesas and mountains appearing ahead, vanishing behind. In the Ice Age, to know about a place was to have been there, or at least heard tell of it. Word of new places came in the form of birds from far away and a different grit to the rivers.

We'd left in a window, a swing season. Too early, and you could be pinned down for weeks in late spring whiteouts. Too late, and snow would start melting off the ice, turning its surface into a web of blue-eyed lakes, waterfalls roaring into widening crevasses. This is when you would have gone in the late Pleistocene, a keyhole to slip through. Some Ice Age summers would have been warm enough that meltwater lakes would be forming and bursting across the face of the ice sheet. But most summers would have been frozen solid, nothing but snow accumulation. If we hit it right, a couple thousand feet above here we would drop our sleds, snap into skis, load gear behind us, and strike off like mariners across the Harding. Doing this would require an advanced base camp, where we could leave extra gear and supplies for a round trip. Ice Age people would have done the same, setting a cache for a return. They may have been crazy, but they weren't idiots.

The first night, we camped on a mountain's shoulder. Clouds filled the valley, but up here we had a clear sky in the long blue light of an Alaskan evening. A glacier poured a hundred feet below us, its blue-bellied crevasses pillared with snow where it came off the icefield. We clinked cups together, celebrating our first load. The two older guys on the trip, Stevenson and his lifelong climbing buddy, John McInerney, who was fifty-nine, had brought Scotch whiskey. They called it the drink of mountaineers. They added clumps of snow instead of ice and raised a toast to a safe and meaningful journey onto the Harding. I toasted back home to my kids and my wife, their pictures carried in the top of my pack.

Later on, after I came home from the Harding, I would take the wall map of North America that was in my office and turn it sideways. I wanted to make sense of what seems to be no pattern at all—just what archaeologists have managed to find from Ice Age arrivals on this side of the land bridge. I tacked up this edgewise continent next to my desk so that every time I glanced at it, part of me forgot what I was seeing. My life was becoming a blizzard of papers and books, desk swamped, plaster cast of a sabertooth cat skull stacked on a pile of folders. With the map sideways, it must have looked as if I were going mad.

I wanted to see not North America, but an unnamed, unknown continent. The thumb of Florida and the long peninsula of Baja seemed like natural geography rather than part of an emblem. Cali-

An Ice Age continent during its glacial maximum

fornia became north, the Eastern Seaboard south, and all the letters and words, names of rivers and states, seemed to fall off like dried leaves.

A drowned land bridge stretched off one corner of the continent almost connecting with distant Siberia, while a second land bridge, still intact, snaked along Mexico where it pinched down to Panama, barely touching the head of South America. This peculiar landmass had rough high country across its upper third, a vast, river-laden plain in the middle, and low mountains followed by a gentle coastal plain on the bottom. I posted it with sticky notes and pinpoint arrows, charting the wide-open spaces between known Paleolithic finds, looking for how people would have moved, what lands they might have crossed.

I'd hold up my hand to form a blank space, an ice cap that covered half of this imaginary continent. Ice was the biggest topographic, environmental feature in the way of the first people, something they had to contend with.

Somehow, people made it over, through, or around this ice sheet long before it retreated, before an inland passage opened between the massive Laurentide Ice Sheet sitting on the Canadian Shield with its neighbor the Cordilleran, grown over the Northern Rockies and the

Coastal Ranges from the Aleutians to Washington State. The two ice sheets had fused together, forming a single mass.

In a hall full of hundreds of colleagues in Santa Fe, New Mexico, I watched archaeologist James Adovasio slam his hand down on the podium and say, "You can die now," his blunt comment aimed at the dwindling number of scientists who insist on a thirteen-thousand-year-old date for human arrival, a holdover theory from the 1930s, back when adventurous bands of mammoth hunters were thought to have marched in from Alaska through an ice-free corridor and then dispersed into the American interior. Good starting place, but human stories are never so simple. Adovasio had been excavating Meadowcroft Rockshelter, a gloomy cave in the wooded hills and outcrops of southwest Pennsylvania, immediately south of the former ice sheet. There, he and his crews found human deposits between fourteen thousand and sixteen thousand years old, and below that were five rough radiocarbon ages from fire hearths and bits of what appear to be carbonized basketry closer to twenty thousand years old, the middle of the last glacial maximum.

Chesapeake Bay, outside Washington, DC, has produced stone tools in earthen layers between seventeen thousand and nineteen thousand years old, a time when this part of the Eastern Seaboard was popular among large carcass scavengers, short-faced bears finishing off fallen mastodons and mammoths. A human-made tool, along with a mastodon skull and tusk, was picked up by a scallop dredge working Chesapeake Bay. The dredge hung up on the skeleton in two hundred fifty feet of water, forty miles off the Virginia Capes, and when the boat captain pulled it loose he came up with bones, tusk, and a lance-shaped blade. The artifact was long and thin, made of Virginia rhyolite, a local rock later determined to be stained with the same chemical patina as the tusk, meaning the artifact and the mastodon had the same history—inundated first with a saltwater marsh, then submerged in mud at the bottom of a bay as sea levels rose. When the mastodon remains were subjected to radiocarbon testing, they revealed a maximum date of twenty-three thousand years old, making this darkly colored blade possibly the oldest human artifact known in North America.

Similarly old dates continue down the Eastern Seaboard. The Cactus Hill site along the Nottoway River on the coastal plain of Sussex

County, Virginia, has produced ages between 18,200 and 22,000 years old for quartzite flakes and blades, and charcoal from fire pits. These dates came from radiocarbon analysis done in the 1990s, which was then corroborated in a 2006 study using optically stimulated luminescence, measuring the last time quartz sediments in layers of sand saw daylight.

Meanwhile, sites from Oregon, Texas, and Florida have produced evidence of strong human presence between fourteen thousand and fifteen thousand years ago, a time ice still blocked the American side of the land bridge.

Continuing down the globe, humans were not latecomers to South America. Instead, human arrival appears to grow older below the equator. One of the key sites in all the Americas is Monte Verde on the coast of Chilean Patagonia, ten thousand miles from the land bridge, where a human encampment is firmly accepted at 14,500 years old. Deeper excavations at Monte Verde have continued to find human-flaked stone down to seventeen thousand years old, atop what appears to be a habitation layer with clay-lined pits from thirty-eight thousand years ago. People were a long way from the land bridge at surprisingly early dates. As if entering a vacuum, they spread to the farthest corners.

Brazil has been weighing in with luminescence-dated twenty-thousand-year-old finds, chipped toolstone from a rockshelter in the canyon-bound Serra da Capivara at the Toca da Tira Peia site. A radiocarbon date came from a mastodon at the Taima-Taima site on the north shores of Venezuela, at least 15,240 years old in association with stemmed projectiles similar to those found at Monte Verde, 3,750 miles to the south. The Taima-Taima projectiles are also similar to those found on the coast of California and the Pacific Northwest, indicating a coastal migration, going around the ice rather than over it. Stemmed projectiles play an important part in the story to come. A stemmed point is a distinctive style associated with people living around the Pacific Rim; the point has a tail, an elongated bottom for securing to a shaft. However it happened, whoever they were, people found a way in during the cold heart of the Ice Age. What would have driven them to attempt it is another question.

She said my name. It was almost a whisper.

I opened my eyes in the tent. Sarah Gilman's sleeping bag was crammed in with mine and a friend named Q, James Q Martin. He was asleep, she was awake. It was our second night camped near the edge of the Harding. The tent glowed faintly in dawn light just after midnight.

"Do you hear that?" Gilman asked.

I listened and heard nothing but the mountain breathing, tent fly wrinkling and stretching in the breeze.

Last time I was awake, she'd been poring through a glacier safety manual, preparing herself for conditions ahead. We called her Safety Sally, a name she went by on wilderness treks, known for being mindful in remote and dangerous terrain.

The book was closed and it seemed as if she'd fallen asleep and been woken by something outside.

She whispered, "Listen . . . listen."

Across our snowy basin I heard a faint warble. Then, another.

"Songbirds," she said. "Thrushes, I think. They've been flying up from the valley."

As I listened, more came through. They were passing in from somewhere else, checking into summer breeding territories. They were the first of the season, and they'd stay through as snow melted, when marmots and foxes would run around the tarns and tundra. I could hear the flutter of their wings, landing here at the tail of a storm, just like we had.

I had asked Gilman why people would have ventured across ice in the late Pleistocene, and the answer she gave me was, "People do shit."

I had to agree, but there must be more. You wouldn't cross eighteen hundred miles of ice just to do shit. On the other hand, they may have known there was another side to the ice, giving them a reason to venture out. Birds may have been the ones telling them the world did not end here, that another world lay beyond the ice.

During the last glacial maximum, glaciers prevented most migratory birds from reaching Alaska. They couldn't cross those wind-stripped and deathly expanses to nest in the north the way they had during interglacial periods. Deflected and forced to breed on more crowded grounds south of the ice sheet, these birds experienced what is known as a genetic bottleneck, evidence that their ranges were reduced, pressure increased, numbers strained, diversity compressed. Glacially fragmented genomes have been found in guillemots, auk-like seabirds, and in dunlins, an Arctic sandpiper. They have also been noted in bears and reindeer trapped on either side of the ice sheet.

Not everything was held back, however. A robin-sized shorebird known as the red knot, *Calidris canutus*, which wintered in South America and summered in the Arctic, continued flying to the north-ernmost parts of its range. Its genes show no sign of discontinuity, which means it was coming across the ice to breed in Beringia.

The same is true for an Arctic-nesting subspecies of sandhill crane, *Grus canadensis tabida*. Every spring, this large and noisy bird would have appeared from over the ice cap. People living on the Alaskan side would have known that across the ice, somewhere far from view, it was coming from habitable country. The cacophony of sandhill cranes arriving every spring, and then raising a ruckus upon depar-ture every autumn, was proof that something lay beyond. Cranes told of unseen lakes and rivers, while the marsh-pecking red knots told of the distant coasts of lakes and ocean. Every fall, people would have watched the birds fly south, and wondered where they were going.

The first people were aware of the signs and signals of the natural world. Their artifacts were projectiles, blades, and ivory sewing nee-dles, either used on animal products, or made from them, or used to procure them. The world around them was a cycle of animals of all sizes, from voles and falcons to some of the largest mammals seen in human evolution. Instead of burning wood for fires, people used megafauna bones, a more abundant resource then. Fire pits as big around as you can reach with your arm, found in central Alaska from fourteen thousand years ago, contain the ash of primarily burned bones, the ground saturated with charred grease. Archaeologist Charles Holmes, who excavated some of these fire pits, tested the technique himself with moose bones. He told me he'd gone through a lot of kindling, but once he got it going, he could keep a megafauna bone fire burning at a steady 400 degrees Fahrenheit, as good if not better than wood.

We, too, were following animals. At least one. Our first day hauling gear out of the valley floor, Gilman found grizzly tracks halfway
up our route to the Harding. The bear was inexplicably heading onto
the icefield, same direction as we were. The tracks were deep, claws
raking through every step, species clear by the length of the claws,
Ursus arctos horribilis. Gilman—a science writer and biologist by
schooling—looked at them scientifically, seeing a context of migrations. Bears are not so different from us: similar diets and habitats,
omnivores, explorers. With recent historic glacial retreats, coastal
grizzlies in Alaska that have been genetically isolated from each other
are coming together. Mountain passes not used for many generations
are opening as the ice melts. Glaciers step back and genes mix, forming a grizzly DNA renaissance.

The grizzly we were incidentally following must have come fresh
out of hibernation, pursuing a scent on the wind. Or maybe it was
like the song, and the bear went over the mountain just to see what it
could see. It was two days ahead of us, at least.

We hadn't expected to find bears, or anything else alive, out here.
We were unarmed, no one willing to take the extra weight. (Ukuleles
are much lighter than firearms.) With our power gels and granola, we
were a walking candy store.

If it came down to a bear confrontation, I'd turn to Gilman, three-
time winner of a women's log-splitting competition in a Colorado
mountain town, the right person to have by your side in a wilderness
like this. We had ice axes, so Gilman was the obvious choice.

She pulled off a glove and reached down, holding her hand over
one of the tracks. Her outspread fingers barely filled the space, each
track unfolding into a big white flower.

If she had approached an icefield via Alaska twenty thousand years
ago, and discovered a line of bear tracks twice the size of our grizzly, those of *Arctodus*, she may have had the same reaction. *Pause.
Think about the animal. Worry a little. Keep going.* She would have
been loaded down with hides, her bundle containing stone preforms,
a ready spearpoint, a small stone blade for repairs, a bigger one for
cutting meat and hide. She would not have come out here to gather
summer wool from willows; she would have left her village on some
kind of mission, and that would have required the right kind of gear.
None of it would have survived. Organic perishables seldom last this
long. Nothing would have remained of her clothing, or the wooden

shafts or handles in her kit. Her fireboard and drill would be long gone.

Bronze Age Ötzi, the mummified fifty-three-hundred-year-old man found on the glacial border of Italy and Austria, carried a flint scraper, an awl, and a small collection of dried mushrooms—a species useful for both starting fires and healing wounds—and his body was found with boot-like leggings made of goat skin, a bear-skin cap, a cape and coat made of woven grass, and shoes like moccasins made from deer and bear leather. He had packed his shoes with moss and grasses to keep his feet warm and padded.

A person from the Ice Age would have had more animal products than Ötzi. Ötzi lived in the Holocene, and the Pleistocene was a different time on Earth. Humans were few, animals many and oversized. In Ötzi's time, there were predators around, but just the ones we know today, no cave bears or sabertooth cats. Ötzi would not have worried about stumbling onto *Arctodus* tracks on the way to an icefield. But his more distant human ancestor, one from the Ice Age, would have scanned the horizon with a fierce sense of alertness, perhaps bringing a pinch of snow up to smell. How fresh? Male or female? Besides a small swallow of fear, this person might have felt hopeful. Unless it was heading out to die, a bear marching onto an ice sheet might have known where it was going, might be leaving a path to follow.

Our camp was set near the edge of the Harding, three thousand feet above the valley where we'd returned for more supplies. Three days in, we were set to climb onto the ice and start a chain of camps, but we were sluggish from the last few days of snowy mountain travel, the elders popping ibuprofen. Clouds would engulf us, then let go, whiteouts followed by blue sky. Waiting for a signal, a reliable clearing in the weather, we rested around a sitting trench we'd dug for a kitchen in the snow. I thought the same scene must have unfolded in the late Pleistocene, an ice party looking for a route, rumors of other lands, birds flying in from who knew where as the people lounged on a half-cloudy day, waiting for a reason to go on.

I plucked a cord on the ukulele, positioning my fingers as the woman at the trailhead had taught us, twanging out of tune. It wasn't music, but it was close. I stopped playing and looked up. The other four were looking up, too. At the crest of a rise behind us, where our tracks came around the edge of a fresh snowslide, two figures were looking down at us in our basin. They were dressed in wind jackets, shorts, and what at first glance appeared to be tennis shoes. They carried only small waist packs and water bottles. A man, a woman. Young Germans, I thought. They must have started hiking early this morning at the trailhead, and when the trail disappeared under snow, they followed our tracks, postholing a few thousand feet up to our camp, which seemed ludicrous without a heroic amount of gear.

For a moment, the world wasn't right. We must have looked like colorful astronauts in moonboots. Our tents were staked into the snow with ice axes. We'd been thinking ourselves as in the middle of nowhere, more adventurous than Ötzi, *National Geographic* material at the very least, and these two visitors blew it. They continued toward us, bunny-hopping from one posthole to the next to reach our camp, as they had to do if they stayed on our tracks. Stepping out, they would have sunk to their knees, and they'd never get out of here. The pair paused when they came close and glanced at us with embarrassment—embarrassed for themselves or us, I couldn't tell. A polite wave, *hi, nice day, have fun,* a couple of questions about their need for gear and watching out for changes in the weather, and they were off, exploring farther along our tracks where we'd been scouting to the icefield.

I was wrong about them being German; they were American. But they were, indeed, wearing tennis shoes. They were unarmed, unspectacular. Their presence didn't make sense to any of us, so we started packing right away.

The causes of migration are many. Something pushes you, something else pulls. This couple was our push. We were not as far out as we wanted to believe. In the vastness of Beringia, it may not have taken much of a crowd to send some people fleeing.

We broke camp swiftly, packs loaded and ready to go. When the couple came back, we were hoisting full loads onto our backs, grunting and tightening straps on each other's packs. They wished us a good trip, and we wished them the same, wandering apart, not sure we ever should have met.

Plates were pushed back at a café table in Anchorage, napkins wad-ded. Five Alaskan archaeologists sat around talking about what they would do if they had a time machine. Everyone gave a different date in the Ice Age, but picked the same area—the Tanana River of east-ern Beringia, within sight of the ice sheet that once broke over the edge of the Alaska Range.

Charles Holmes, who was from the University of Alaska at Fair-banks, had discovered and excavated a handful of Ice Age habitations and mammoth-butchering sites along the Tanana. He'd been work-ing on a knoll called Swan Point, a 14,500-year-old hunting camp where tools were fashioned, weapons sharpened and repaired, read-ied for the hunt. Trim, gray-bearded, Holmes said, "People would come together to do roundups. It would happen once or twice a year. We've got a fifteen-year-old female mammoth, a five-year-old juve-nile, and a baby all together. I can envision this little group of mam-moths getting whacked, not necessarily right on the site but nearby, and then processed."

Holmes said, "What we have are the first human footsteps in Alaska." Swan Point has been considered the oldest bona fide archae-ological site in the state, not just a butchered Pleistocene bone but a real settlement. Someone at the table called him out, saying Swan Point couldn't be the actual first settlement, and Holmes corrected himself. "Clearly, if they're in the middle of the state, they aren't the first." Holmes used "first" the way most would, thousands of years reduced to one word.

Sitting across from Holmes was Diane Hanson, from the Univer-sity of Alaska in Anchorage. She said, "It was a window of oppor-tunity fourteen thousand years ago, a hunter's paradise. Before that, you would have headed south or to the coast, but you wouldn't have stayed here."

Hanson was referring to the last glacial maximum. A glacial max-imum is the coldest it gets. Most are about 10 degrees Fahrenheit colder than the North as we know it today, enough to push ice caps to their greatest extent, places like southern Wisconsin turned into moraines of rock rubble, indicating that this is as big as the ice grew.

The height of the last glacial maximum was about twenty thousand years ago, around the time people were showing up, based on some of the early dates from Bluefish Caves. A massive ice sheet was sending out frigid winds, freeze-burning the land around it. The land bridge might have been a survivable place at the time, while the Alaskan interior would have been a suffering and hungry country, not like the heyday of Swan Point 14,500 years ago, but like the barren edge of an expanding ice sheet. Layers of coarse, wind-blown sediment have been found throughout the Alaskan interior during this maximum, indicating a lot of blowing grit, while sites across Siberia were being abandoned, the people heading south. They had to find refuge, perhaps following the Porcupine River to Old Crow Basin, nestling into a megafauna sanctuary around Bluefish Caves. Wherever they went, whoever they were, there were few of them, knocked down by generations of long, deep cold, not leaving much, if anything, for archaeologists to find.

"Where are they, then?" asked David Yesner, an imposing figure at the table. "Why aren't we seeing them?"

Yesner asked this before the 2017 revelation of a twenty-four-thousand-year-old Beringian horse jaw with butchering marks from Bluefish Caves, a single find that is still not enough to convince many archaeologists. Yesner, with Brillo-pad hair and rings on his fingers, is an icon in Alaskan Paleo sciences. It's hard to disagree with him.

Another archaeologist at the table said, "It might be a problem of demographics."

Leaning back in his chair like a doubtful king, Yesner said, "It's because they weren't here."

The last glacial maximum is a hard time to find anyone in the Far North, especially on the American side. Hanson works the Aleutian Islands, an arc of old and some new volcanoes reaching six hundred miles from Alaska almost to Kamchatka. "Between sea-level changes, glacial rebound, and tectonic uplift and subsidence, it's hard to figure out where things were," she had told me. "Some places are up in the air just in a few thousand years."

The boat taking Hanson and her crew to their excavation sites and surveys costs $10,000 a day, and sits offshore waiting while they work. "There are four archaeologists working on a thousand-mile-long island chain," she said. "You can see why we can't find anything."

It's not true that she hasn't found anything. Hanson has been mapping three-thousand-year-old villages and camps, but so far nothing as old as the Ice Age. Information that far back is spotty, the record shot through with more holes than data. The poor quality of what is known has scientists pounding podiums, some threatening libel suits over contradictory papers. How, when, and where the first people entered the Americas remains one of the world's most contentious prehistoric problems.

The climb onto the Harding took us up a steep, blank canvas. No tracks lay ahead; the last storm had abolished all human presence. It was like being born, or maybe like dying. I'd all but forgotten the visitors in tennis shoes. I tried to forget everything, the cities, the deadlines, all of it.

A new storm was moving in, summits disappearing around us as a light snow began. I sensed when we passed from snow-on-mountain to snow-on-ice. Gilman was marching ahead of me, her pack loaded with skis, ropes, a sled, and an ice axe. She had started to sing. Something must have shifted beneath us. I couldn't hear the words, just her voice as she kicked in steps, snow to her shins. She told me, later, that it was her fear of crevasses. She'd felt them getting closer, like circling sharks, and so she sang to ease the hike, remembering all those she loved in case one of her steps crashed through and she disappeared into a blue hole below. When she started singing, I figured we must have come onto the Harding, her lullaby carrying us into the storm.

By morning, our world was gone. The icefield and our camp had been swallowed by an unbroken whiteout. We should have been able to see a corridor of nunataks sticking up through the ice. Instead, visibility had been reduced to near zero. It wasn't eyelash-freezing cold. The wind was gentle, snow steady but not enough to bury us anytime soon. The sky had simply come down and smothered us.

Snow slid off tents as we opened zippers. Heads peeked out. The sun was up, but who could say where? We appeared to be floating in a blue-gray emptiness, as mountaineers describe it, the inside of a Ping-Pong ball. Our window had closed.

The first man out was Stevenson, stomping into his boots and squinting through the snow globe around camp, seeing nothing. His partner in the tent, John McInerney, had been climbing with him since they were in high school. McInerney said, "Condition report."

"Conditions could be much worse," said Stevenson.

"Much worse? Like raining molten lava?"

The air tasted like ice. I couldn't get anything more out of it. This wasn't traveling weather, that much we agreed on. Heading out blind across unknown snow bridges and the blocky upheavals around the bigger crevasses was ill-advised when we couldn't see a thing.

For that first day of the whiteout, we were either out in the nothingness, or cooped up in tents. We all had to do something. Gilman buried herself in Dostoevsky. Q, our photographer/filmmaker, checked batteries, cleaned lenses, cleared memory cards. We three had one tent. Stevenson and McInerney were in the other, the two close friends whiling the hours reading thin paperbacks, as old-school mountaineers are wont to do.

I mostly stood at the edge of camp and stared at blowing nothing, waiting for anything to emerge. To my back were the vanishing crisscross of our boot prints and the bright blossoms of tents. We may think of our far-ranging ancestors as stoic and unquestioning, their senses raised majestically to the wind, but I can imagine one of their band stopping about now and asking, "Where the hell are we going?"

When I turned back to camp, Gilman was sitting in the entry to our tent. She was a willow-limbed woman, youngest in the group, early thirties, tall, with short dark hair dyed flame-red at the tips. She wore a cone-shaped woolen Soviet hat, red star emblazoned on the front, that made her look like a Russian jester. The rainfly was unzipped, and she sat half in, half out, sketchpad balanced on her knee. Trained as an artist and schooled as a naturalist, she has a sharp eye for landscapes. She said she'd made a drawing of me and asked if I wanted to see it. I wasn't sure. I was feeling out of place in my ice costume. She turned the sketchpad toward me.

It looked like a joke at first, a blank page. Then, I saw a lone figure in a lower corner. I was as small as a grain of rice. With pen and ink, she'd captured the way I stand, with my weight on one leg, hood pulled up in the wind. The rest of the page devoured me.

She said, "This is what we look like out here."

Beringians—I saw them in skins and furs, hunkering down as storms howled over them. They closed their eyes and waited, snow building up. Hunting parties gone astray, or explorers testing the boundaries of the world, you'd expect many to have perished in sudden storms, or been eaten by giant bears on the ice, their spears swatted away like toothpicks. The bones of these lost souls, and their last withered skins, would have drifted into the ice, been swallowed by crevasses, and turned into a soft powder that rivers would eventually carry to the sea. They'd never be found.

For two days we drifted inside a place with no up or down, the sun passing invisibly through the sky. Shadows were gone from our faces. Q and I were antsy, too much time cramped in a tent that smelled of fouled synthetic clothing. We came up with a scouting mission. We'd go out as a pair a short distance, half an hour. With the lay of the terrain, Q felt safe; the surface didn't have the right bend for crevasses. We'd done rock and ice in South America together, and I trusted him. We decided to head out with a GPS, marking camp, going out as far as felt comfortable and then coming back. On the sled, we had extra gear and rope in case the day went sour. As we snapped into skis, one of the elders pulled a zipper open and stuck his head out of a tent. "If your GPS fails, don't try finding your way back," he said. "Wait it out."

"Got it," I said.

Q punched a waypoint into the device and said, "We'll be careful."

You see an old friend smile like this and think, some Paleolithic explorer probably beamed with the same expression. *Let's see how far we can go.*

The popular version of the first American colonization is "people on their way here"—like immigrants on their way to Ellis Island, bustling to see ahead. But people weren't on their way to North America, because nobody even knew North America existed, and so their arrival was an exploration every foot of the way. There might have been adventurers—crazies at the edge of all human societies, willing to do nearly anything, wild-eyed characters heading always for the hills—but most people would have spent their time surviving in hungry country, weathering the unstable ups and downs of worldwide glaciation, in the company of several tons of animals per square kilometer. Life was hard enough, and short enough, already. Why go off on a lark across the ice when there was already so much to do where you were?

The drive may be innate. Some fruit flies, when they're young, are adventurers. When they first chew out of their pinhead eggs as larvae, some wriggle farther than the others. Scientists have studied these individuals and identified a genetic locus connected to foraging, what is known as a *rover allele*. When the presence of rover alleles is increased, a larva will pass up food and creep toward the limits of its petri dish. Such larvae tend to be found dead around the edge of the dish, their journey more important than living.

For a human, this might be identified as a calling, an urgent voice. Or insanity. The increased genetic presence of the dopamine receptor known as D4 is correlated with restless behavior and what is known as "novelty-seeking"—the kind of people who are reckless or adventurous, in need of something new. Perhaps this is why Q and I took off into a whiteout, or why any of us were here in the first place. We are the tiny maggots you'd find around the edge of the petri dish. In a book of short stories he wrote about mountaineering, Stevenson called the urge to summit "a trance-like mystification, feeling privy to some large but unintelligible secret." When I asked McInerney why he was here, he said, "When the hell else am I going to do something like this?" He told me that when his colleagues stand around the water cooler back in DC talking about their vacations, they can scarcely conceive of what he does. He added, "I barely understand it myself."

A genetic study of more than two thousand prehistoric individuals worldwide, ranging between one thousand to thirty thousand years old, found that this pronounced D4 marker is more prevalent among those who migrated as compared to those who maintained a long genetic history in one place. Among Native American genomes and those of their ancestors, the presence of D4 is correlated with an individual's distance from the land bridge. North America, with the closest access to the land bridge, shows 32 percent of samples with D4 elongation. Central America comes in ahead with 42 percent, and South America reaches an average 69 percent, as if people needed that much more *umph* to reach that far south.

Too high in D4, though, you'd never be seen again, a seed blown beyond all horizons.

The Harding is a tricky spot to get through. Compasses are unreliable. A mineral anomaly in the rock buried hundreds of feet below taps the needle off-course enough to cause problems.

If you hoped to use your innate sense of direction, you'd have to surrender out here. Some living organisms use what is called *magnetoreception*, a sense that enables birds, lizards, and insects to know what direction they are heading. Crystals of the mineral magnetite have been found in human brain tissue, orienting to cardinal directions. Our brains are compasses. Whether or not we use them, or are even aware of them, is a personal issue. But on this icefield, using the body as a compass would be useless. In the blind of snow and cloud, we were relegated, finally, to talking to satellites.

Tents vanished from sight after forty feet, voices taking longer, the mumble of conversations fading after a hundred. But who was counting? I paid attention to the strokes of skis and poles, feeling as if we'd fallen into space. Snow picked up in the wind. Grains hissed across each other. Instead of skiing single-file, Q and I traveled side by side, our senses flung wide like a net. My eyes panned, strained, any block of ice sticking up, any rise or fall. Nothing. The air cooled, snow came on harder. Crystals turned from woolly bars to cold six-pointed flakes near the edge of our petri dish.

Q stopped to look back. I stopped, too. Our tracks vanished in the gray, blowing murk.

"I think we're hooking left," he said through the wind.

I said, "I thought we were hooking right."

It had been an hour. I pulled the GPS out of Q's pack and handed it to him. Holding it in the air, waving at satellites, he punched buttons until the device picked up enough signal to pinpoint a location. We'd come about a mile from camp and were 90 degrees off course, heading up a gradual incline that would lead into a field of mountains, not a place one should go in a storm.

He punched in our route home. A mile later, our tents emerged from the emptiness. They looked like lifeboats coming out of the fog, bright and strange to see.

The human brain in the late Pleistocene was 5 percent larger than our modern brains. What we have lost since then is hard, if not impossible, to tell. Only the decrease in volume is evident, not what is missing. Before you think of Pleistocene people as potentially smarter than us, remember that volume does not necessarily translate to function. Domestic animals have smaller brains than their wild counterparts, a result of atrophied aggression, not necessarily reduced intelligence. This is what allows them to live in tighter quarters without ripping each other to shreds. More compact regions of the brain develop to address fealty to unseen forces, a bond between individuals taking up less space than the need to kill one another.

More information may have been coming in during the Ice Age, a constant flow from the outside world, earth, wind, sky, valley, mountain, beast. Neuroscientist John Allman wrote, "Brains exist because the distribution of resources necessary for survival and the hazards that threaten survival vary in space and time." He sees little need for a nervous system if the organism lived in "regular and predictable surroundings." It is the complexity of the world that makes us what we are. "Brains," Allman continued, "are buffers against environmental variability."

A bigger brain might be a bigger buffer. Environmental variability for a Paleolithic human would have been significantly greater than for most of us living at around 72 degrees Fahrenheit with smooth floors and ergonomic chairs, or any chairs, for that matter.

The *angular gyrus* is a region located in the back of the head, key in semantic processing, reading and comprehending words, numbers, retrieval memory, spatial cognition, reasoning, and social cognition. This is probably where a lot of our mass went. Not much larger than a couple grapes, the angular gyrus may have replaced the Paleolithic 5 percent.

People who came to the land bridge were making ornaments of shell, bone, and teeth. Human figurines twenty thousand years old in Siberia were carved from mammoth ivory showing hoods, bracelets, shoes, bags, and gear on their backs. Skin-and-pole huts were erected and held down by rings of mammoth bones or caribou antlers. People came with languages, the complexity of their speech near or equal to ours. Cooperative hunting would have required communication between hunters and their domesticated wolf-dogs, where multiple juvenile and adult female mammoths have been found butchered side by side. This would have required communication, a game plan.

Language doesn't fossilize, but it does appear with a host of artifacts, musical instruments, depictive artwork, and the spread of exacting tool and weapon technologies. The late Soviet cognitive scientist Lev Vygotsky wrote that these very human expressions "extend the operation of memory beyond the biological dimensions of the human nervous system and permit it to incorporate artificial or self-generated stimuli, which we call *signs*."

Beringians would have most likely spoken a proto-Indo-Eurasian language. Anyone venturing across the Laurentide and Cordilleran ice sheets would have quickly developed words for *crevasse* or *ice heave* or *nunatak*, signs on the way ahead.

When the storm broke at the end of the second day, a curtain was drawn back. The massive breadth of the icefield came suddenly into view. Dopamine receptors stood on end as clouds unfastened from each other, the silken expanse open, showing not a track or blemish. Q and Stevenson sat at the kitchen trench we'd dug. Gilman stood next to a cluster of skis planted near the tents. McInerney unzipped his tent door and looked out, seeing rags of sun-shot clouds speeding below a topaz sky.

As the hole widened, nunataks began to appear, summits in the distance coming out of nothing. Our eyes weren't sure what to do with it all. The world seemed to quaver. Q let out a whoop. But Gilman folded her arms and squared her jaw at the rest of us. She asked, *What if it's a sucker hole?*

Taller than any man on this trip, she was immovable. When she had said earlier she wanted for all of us to practice self-rescue knots, McInerney had grumbled under his breath, "Chances are if anyone falls through a crevasse, they're just fucked, they're not coming out alive even if we are roped up." She'd been expressing concern over the past day and a half: not enough attention was given to safety plans, not enough time scouting and surveying what could certainly be a dangerous approach. Every trip needs someone to be saying, *Do we know what we're doing?*

"I am uncomfortable leaving right now," Gilman said, pointing out that it was after seven o'clock and we only had a few hours of good daylight left. If the hole closed back in, she wanted to be at *this* camp, not strung out somewhere on the ice.

Q, our other strongest member, a short, burly figure who could carry any one of us out of here over his shoulder, called bullshit. He, more than anyone, wanted to get out there. Nunataks were waiting, and he wanted to see Stevenson climb one. Q said that if we stayed, food would run out. Fuel would be used up for melting snow. We'd be sitting on our asses, our trip basically over, and we'd have gotten nowhere.

On any good journey, you'd need both a Q and a Gilman. One pushes the trek to go on against the odds, and the other makes sure you don't die.

"Let's give it an hour," said Stevenson, another needed voice, a good leader to smooth it all out and make the call. He said, "We can have some food, fuel ourselves up. If the window's still open, we go."

In an hour, the hole in the storm was wider, a hallway of nunataks revealed across the blank face of the Harding. Q and Gilman, like a pair of siblings, were done with their feud and had gone back to chest bumping.

We took camp down and moved gear into sleds and backpacks, clicking into our ski bindings to get as much distance as possible before the day ended. We set a line across the expanse, clean wind racing around us. Q broke out front, a raw shot of energy on the icefield, sled hissing behind him as he parted from the rest of us as if forgetting he'd come with anyone. He lifted a pole in the air and let out a Viking holler, nunataks littered ahead like asteroids.

At a sunrise camp on the icefield, our tents were frosted like two colorful eggs. We'd established ourselves in a miles-wide hall of summits, a statuary of ragged nunataks. Dark surfaces of metamorphic bedrock were melting through, the mountains mostly white with snow.

We left camp for the day, skiing into snowy rises. At the first ridge, Gilman and McInerney decided to hang back. Gilman didn't like the cavernous threat of bergschrunds, the deep crevasses that form where glaciers pull away from mountains. She said she'd rather be sketching. McInerney smiled and waved, saying he'd had enough adventure and did not need to go up one of these nunataks. That left Q, Stevenson, and me to summit.

We contoured on skis into a group of rugged peaks standing around each other like a coven. The simplest peak was our aim, a pyramid with a rock-heap summit broken up through snow cornice, one final piece of land the glaciers hadn't been able to eat. In a stadium-sized bowl on its flank, fumbling with gloves to tighten one of his ski bindings, Q unclipped his ski from its leash, and it slipped out of his grip. He dove, planting the side of his face in the snow, arms thrown out. He missed the ski by inches. It flew. Without a word, Stevenson drove a pole and turned downslope, slicing after the solo ski.

Stevenson had told me there's only one mountaineering rule you really need to know: "Don't fuck up." Small mistakes add up out here. As Stevenson and the ski lost elevation, the scope of this place began to resolve. The mountain became bigger, the sweep of snow rising and falling between other mountains onto the face of the Harding, which grew by every hundred feet of descent. Q cursed as he got back up, watching Stevenson and the ski becoming smaller.

Later, Stevenson said it was one of the best runs of his life. The ski followed a perfect line, path of least resistance. He stayed on it at top speed like a shadow. When it planed out a mile below onto the icefield, pursuer and pursued finally stopped and we could see where we were, the scale of the earth revealed. We'd lost an hour. Stevenson was exhilarated, beaming, when Q went down after him, hobbling on one ski to catch up with his other.

Saying little, our lips chunky with sunscreen and flakes of sun-cooked skin, we ditched our skis when we reached the place where snow on ice turned back into snow on rock. The summit was a few hundred feet above us. Hand over hand we climbed in stubborn, hard boots up a steep angle of repose, staying out of each other's lines, avoiding rockfall and crystalline cascades of corn snow. Crab-shaped squalls flustered around us, taking other summits and giving them back. The ice sheet below was a lake of cloud shadows, our camp so far out it hurt the eyes to try and find it, like looking for a hawk circling so high it disappears. As we reached a swayback saddle buried in snow near the top, Stevenson and I turned for the summit. Q went the other way to set up the tripod he'd been carrying. He wanted to film Stevenson's triumphal arrival atop his first nunatak.

The two of us scrambled up a shattered, blocky summit, the top barely large enough to stand on with two boots. Stevenson went first. He secured both hard soles and he rose up, lifting his arms—top of the world, king of the hill.

Did the first people summit so triumphantly? Attaining such a height would have been like unrolling a map, seeing where you are. Birds may have told of a continent ahead, but every horizon offered

Raven and the nunataks

many choices and turns in direction. Who knew how much ice was between here and the other side? At least Scott and Amundsen knew there was a South Pole, and how far away it was, what provisions were needed. What did Ice Age travelers see?

I see a chain of stepping-stones across the ice, one viewpoint to the next. This nunatak route would have descended the Cordilleran Ice Sheet, peaks of the Rockies and the Coastal Ranges sticking up along the continent's western margin. Scraping lichens off the rock and eating birds, a scouting party might have made it a couple thousand miles, discovering an uninhabited continent on the other side.

As we came down from the summit, I sank in snow, working my way onto the bulb of a cornice, a good vantage across the icefield. Every step was a posthole breaking to my knees. Crevasses opened into chasms hundreds of feet below. Beyond those, I could see the tips of other, nearby mountains. Farther out, twenty miles away, a cluster of nunataks gathered like an island before the icefield dipped over the horizon, line of sight gone.

Near the edge of the cornice, I found fresh bird tracks. Where the bird had landed, wing prints touched snow. It had come from the same direction as us, perhaps flying the same route that brought the bear, the same green valley and river behind us where we'd started. It had landed and walked across the cornice to look down the other side, exactly as I was doing.

We think of ourselves as different from other animals. We extol our own tool use, congratulate our sentience, but our needs are the same. We are creatures on a planet looking for a way ahead. Why do we like vistas? Why are pullouts drawn on the sides of highways, signs with arrows showing where to stand for the best view? The love for the panorama comes from memory, the earliest form of cartography, a sense of location. Little feels better than knowing where you are, and having a reason to be there.

The bird was a raven, its toe prints armored and tucked together so it wouldn't sink as it walked. It had lost a dime-sized feather, black around the tip, white at the quill. The feather had melted slightly into the snow, leaving an "X" on the map.

At the round tip of the cornice, the raven had opened its wings again, continuing in the same direction. It might have stopped here for a breather, a chance to rest. Where the last tracks ended and the raven flew off, I could see its black wings spreading. The land took shape as it launched over the ice sheet, the bird becoming smaller, the mountains and ice growing. My mind's eye watched its path continue, a perfect line, a direction of travel that passed through me like an arrow.

A light wind coursed across the icefield that evening. Galaxies of snow squalls drifted left to right. The nunatak we'd climbed earlier stood miles away, appearing and disappearing through clouds. As stoves melted snow, we heard a distant drone. It was an engine. Two engines, their pitches vibrating against each other. Gathered around a kitchen trench, we peered through low, golden light. Two planes cruised the horizon. They had the high-pitched drone of Supercubs, small single-engine monoplanes, room in the cockpit for a pilot and a moose dressed out from a hunt. A couple hundred feet off the deck, the planes swung between squalls like gliding birds. I stood from my spot in the trench, Gilman stood from hers, and Q began unbuckling his camera. Stevenson and McInerney remained seated, whiskey in their cups.

One plane was on the other's tail and just slightly to the side. The formation was a way of communication, two animals crossing the ice, one looking past the other's shoulder. They must have been out for the golden light of sunset, taking in the amazing breadth of white. When they spotted us, they turned in tandem. How could they not? Like seeks like. The sound of their approach grew into a clamor. Stevenson stood, cup in hand.

The two planes dropped just above the deck. At fifty or sixty miles per hour, they did in a few minutes what had taken us six days. This was evolution. We could see a pilot in each window, wings tipping hello as they skimmed past. I launched a fist of solidarity. Gilman reared back with a wild howl. Q tracked them with his camera as if they were UFOs. Stevenson and McInerney turned to watch them go. We were of the same tribe, a species feeling its edges, remembering what it is like to see the world unknown. Same as the couple who'd followed our posthole tracks three thousand feet up from the valley floor, we were drawn to immensity. Something beyond us calls.

The sound of 150-horsepower engines buzzed into the distance and faded. Stevenson sat back down, boots in the trench.

We need each other. One of the reasons for migration is the bond of society, an urge to be together. A group might keep going for no other reason than the hope to find others. Seeing planes out here was a curious relief. They were like us.

After the elders went to their tents, snuggling up with their paperbacks, the three of us were left outside staring into the blue drift of evening storms. The Supercubs had left a delicious wake of emptiness. It felt as if the planes had unzipped the sky, the world even larger than we knew.

I carried in my gear a small amount of red ochre. I'd been traveling with this plug of mineral in a plastic bag in a zippered compartment for nearly twenty years. Ernst Wreschner, a paleoanthropologist at the University of Haifa in Israel, wrote, "Prehistory has produced evidence for two meaningful regularities in human evolution: tool making and the collection and use of ochre." Over time my ochre had decreased, pulled out and used for special occasions every several years. I told Q and Sarah that I had some in my pack, and this was just the night, a celebration of an excellent day, a nunatak climbed, Supercubs for fireworks to close it out.

Digging into the top zipper through pens, twine, knives, bandages, and toiletries, I found the thumb-sized package, a plastic bag twisted around itself and cinched with a rubber band. The red of the ochre was leaking out of the bag, lightly polishing everything in the compartment with a ruby blush on my pens and headlamp.

We asked Stevenson and McInerney if they wanted to partake. A book page turned inside the tent and McInerney said, "You kids go right ahead."

When I'd first collected the ochre, I hadn't known what it was, not technically. It was just a shiny red sediment that rubbed onto skin like paint. I liked it, so I took some, needling it out with a knife tip from a layer of Bright Angel shale in the Grand Canyon. It went almost everywhere with me.

This spectral, blood-red mineral, also known as *hematite* or *iron oxide,* is the ceremonial stone of our species. Its use dates back to the beginning of *Homo sapiens,* found in rituals, human burials, and the earliest art from Africa, Indonesia, and the Iberian Peninsula. A Russian archaeologist who excavated a thirty-six-thousand-year-old grave in Siberia found the pit drenched in red ochre, what he described "as bright as fire and as scarlet as blood."

This is perhaps the most telling artifact to come over with the first people. The use of red ochre in Ice Age America is considered to be a sign of a direct relationship with Old World Upper Paleolithic complexes.

The mineral is utilitarian. Added to pine mastic, it would have helped attach stone tools to their wooden handles. It has preservative properties for skins and hides, and is known to heal wounds. It can be used to tan animal hides and keep off rot, and it maintains the skin of corpses. One archaeologist suggested that its use may indicate "a continuation of efforts to save the person's life."

It also can be painted on skin in a decorative fashion. That's why I picked it out of the rock formation. I knew nothing about ochre's archaeological history at the time; I just liked the pigment it made. Gilman put out her hand and I crushed a few pinches into her palm. She mixed the powder with water, dipped her fingers and swept the color along her cheekbones, up her chin, and down her smile lines. She held the most serious expression, her markings standing out against a backdrop of ice and barren, snowbound mountains.

She took another swipe from her palm and painted my own wind-dried face with a two-finger streak under each eye and three down my forehead. She turned to Q and gave him a set of war stripes, which he took still-faced, as if anointed.

Arizona ochre dried and pulled on our skin. Gilman spread the remaining pigment in her palm and painted out to the tips of her fingers. When her hand was entirely red, she reached down and planted it firmly in the snow. When she lifted her hand away, she left a five-fingered blood mark on the face of the icefield.

4

THE LONG COAST

The inside of the boat looked like bones, as if we were standing in the rib cage of an animal. Turned upside down and raised on blocks, this *umiaq*—the traditional skin boat of Indigenous Northern sea hunters—was made of walrus skins stitched together around a wood frame, eyelets cut through the inch-thick hide and secured with rope.

My mom and I took shelter underneath it on St. Lawrence Island in the Bering Sea. We were returning from a stormy hike along the coast, sleet leaking into the seams of our coats. We hid from the wind and blowing sea waves behind a stack of cargo containers. Then we hurried to the shelter of several *umiat*—plural for *umiaq*—turned upside-down on blocks and wooden stilts at the edge of the village.

We stood up in the protection of one of these boats, pulling back our hoods, and found ourselves inside an animal.

Kayaks and umiat are thought of as living creatures among Northern cultures. You have to treat them with respect, as you would anything alive, especially anything larger than you. My mom reached up and touched the boat's keel above our heads. Her fingers traced the wood. She was a furniture builder who worked alone in her shop, spending days and nights with wood, falling asleep with sawdust in her hair. She questioned the umiaq with her hands. The keel, she pointed out, wasn't cut with a saw, but split out of a larger piece of wood using wedges and pries. The process is called *riving*, taking the

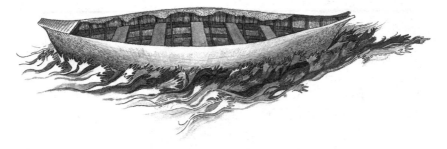

The skin boat, *umiaq*

strongest piece out of a tree. Where the tree came from on this barren island we didn't know. It was prized.

Hammered wooden pegs were used instead of nails or screws. She felt along the frame over her head, her hands bony from years of woodworking, a couple fingers cut down to the first knuckle from a table saw. She thought of the hands that made this boat, seeing them grip each rib, bending it into shape, lashing walrus hide to wooden thwarts. The boat wasn't a piece of furniture to her, not like a table or bench that she might build. It looked more like a basket with ribs and cross ribs secured by fine, strong twine. The wood and hide boat is a universal way of getting around on the water, like the coracles of early Britain, ox hides stretched around large wooden baskets. According to medieval accounts, the sixth-century Irish monk St. Brendan traveled to the East Coast of North America and back in a *curragh*, a long early Irish boat, wood-framed with skins and a sail.

The umiaq is another version of the skin boat. Its roots reach into the early Holocene, and probably long before. People in the Ice Age knew about boat travel. Artifacts from forty-eight thousand to thirty thousand years old have been found on islands far offshore along the Asian side of the Pacific Rim. The only way anyone could have reached these sites was by boat. Boats similar to the umiaq, but made of grass or skins, were likely used, as if the people built sea animals for themselves and then climbed inside.

To the Inuit of Greenland, the umiaq is known as the women's boat, originally paddled by women, while men used skin kayaks, sometimes aided by sails. A kayak is closed, but an umiaq is open and can carry about ten people. It's the family boat. Turned over, they were used as shelters, arranged in a circle or a crescent to form a kind of village. Groups traveled together, kayaks and umiaq moving from camp to camp. They followed hunting patterns and bird migrations, collecting eggs, seaweed, and driftwood, hunting caribou, salmon, seal, and whale. This was before permanent villages, an older way of living, people born into motion.

We'd heard stories of skin boats with families or hunters who went down in foul weather off the coast of St. Lawrence Island. Shores were marked with white wooden crosses, places where people had perished. When people described these disasters, how the waves swamped one skin boat after another, I pictured bodies sinking like angels, seal skins and coats waving their arms as they descended into the Bering Sea.

When I was a kid, our umiaq was a Volkswagen bug, a round shell of a car painted baby blue, plying the dry spaces of Arizona and Colorado. Our family was tight, a single mom and her only child. We moved frequently, sailing across the Southwest like mariners. We would not have survived in the Paleolithic, the two of us alone; we'd have been eaten by sabertooths before we knew it. A family of only two would have meant something had gone terribly wrong, a vulnerable configuration.

The last time I saw that car I was four years old. I was asleep one morning in the back seat near Nutrioso, Arizona. Maybe my mom nodded off at the wheel, or sunrise momentarily blinded her through the bug guts on the windshield. A tire caught the edge of the asphalt, and we flipped. The Volkswagen crashed through a wood-pole fence and rolled a few times into a field. My mom's big, white Samsonite suitcase went flying. I remember how it seemed to pause for a moment in midair just before it landed and pummeled me, and I pummeled it back.

Our baby-blue umiaq came to rest on its roof. My mom and I were fine, no breaks or blood or twisted metal. I remember the pecan pie: It fell out of the upside-down vehicle onto the dry pine-needled ground, and a dog came up and ate it. We caught a ride into town in a postal truck. Rain had just started to fall. The Arizona highlands smelled cool and crisp with pines as windshield wipers pushed rainwater back and forth.

This is from my own origin story. It is how I know who I am. Going ahead without stories or family is like falling off the edge of the earth. When my mom and I sheltered under the umiaq on an island in the

Bering Sea, it was another tale added to the stack, a trail of bread crumbs marking my passage into the New World.

In Native tales from the Northwest, Raven the trickster-god is said to have been walking along the beach one day when he found the first people inside a sealed clamshell washed up from who knows where. He paused, lowering his glistening black head to listen. He heard movement inside, a kind of bumping, like elbows and knees knocking against each other. Curious, he pried the shell open with his beak. From the dark hold emerged the first people, naked and excited. Raven stepped back as they scampered away, the prints of their grape-sized toes heading off in all directions. The world would never be the same.

Who knows how many firsts there were? How many clamshells landed on this side of the world?

The Fukushima Earthquake hit the coast of Japan on April 11, 2011. The following tsunami washed nearly five million tons of debris, including boats, cars, parts of houses and factories, and some of the twenty thousand dead, into the Pacific. Debris entered the gyratory Kuroshio Current and was swept toward North America, where objects began to arrive a month later.

The first landing was April 18, when a cargo box from Yamamoto in Miyagi prefecture, the hardest hit from the quake, washed onto a heavily forested island of Haida Gwaii on the west coast of British Columbia. The box opened as it hit the beach and a Harley-Davidson motorcycle rolled onto the sand. This was the first offering from Japan.

The coast of Haida Gwaii is the same place where the original clamshell is said to have washed ashore. Tens of thousands of years ago it was a clamshell, and now it is a Harley-Davidson, taking a month under no power but wind and sea to make the four thousand miles from Japan to the edge of North America.

Ten to a boat of thick sea-cow skins blown off course, they would have eventually given up trying to paddle home, letting the sea take them. Their drinking water would have been rain or maybe captured pieces of icebergs. They ate what they could spear or hook. Perhaps they prayed to an Ice Age iteration of *Okikurum*, the creator god of the prehistoric Ainu, descendants of the sea-faring Jomon who hunted waters of the North Asian Pacific Rim for more than ten thousand years. The only solace may have been a whale spout. Seeing one and then another, they may have begun to paddle in the direction of the whales, a spring migration to distant kelp forests. A bird would have been spotted, then another. Seals would have lifted their heads from the water. Land would have appeared.

The first footstep onto North America may have been people jumping into the water, dragging their skin boat onto unfamiliar shores. Seeing an enormous ice cap rising above a ragged coast, they would have all asked the same question: *Where are we?*

You would have needed families, not just a lost few. Without grandparents and children, it's not a migration. It's just a lark, as Gilman put it, people doing shit.

With my own family, my two kids and my wife, our clamshell landed where glaciers and mountains fall into the waters of Prince William Sound, south-central Alaska. It was late June. Earlier that month, I'd been on the Harding Icefield. Switching out of skis, boots, and sleds, I was now setting off for a twelve-hundred-mile journey south from Prince William Sound by kayak, plane, and ferry to the upper corner of Washington, where the first ice-free interior would have opened for Ice Age boat travelers. The ice and our nunatak route had been exhilarating, but it was no way to move a family.

Twenty thousand years ago, this coast was nothing. Where the expanding Cordilleran Ice Sheet came to the water, half-mile-high towers of ice plunged into the sea. Places to land were hard to find; most of the shoreline was covered in glaciers. But by seventeen thousand years ago, a coastline began to appear. The coldest times were over, the last glacial maximum done. The first land exposed by warming temperatures and the retreat of the North American ice cap was not inland, but here. The Northwest Coast became a cheat, a way to get around the ice.

Archaeologist Knut Fladmark proposed in 1979 that a coastal route, not the ice-free land bridge, brought the first people to America. In an *American Antiquity* article titled "Alternate Migration Corridors for Early Man in North America," he wrote about the islands and coasts that might have existed then. He noted, "These would have been more environmentally suitable for human occupation than any interior ice-free areas, and marine littoral resources would have provided a relatively abundant living for any people possessing simple watercraft." Biologists in the 1960s had found evidence of plant life on the Ice Age margins of Alaska and British Columbia, a region previously thought to have been ice-scoured and devoid of life. Bear bones from the last glacial maximum were discovered in caves in British Columbia, indicating that isolated populations of large animals survived. There had been refuges, places to land the whole time, a perfect way into North America if you had a boat. If plants and bears could live, Fladmark concluded, then there could have been people.

In 1997, archaeologists and marine ecologists presented the Kelp Highway Hypothesis. The coast, they said, had actually been highly productive during the Ice Age. Kelp is a large seaweed in the order *Laminariales* that grows in towering columns anchored to the sea floor, suspended by gas-filled bladders. Waving through the water, it creates its own ecosystem, from isopods to otters and orcas, one that continues up the rivers and onto the land, where brown bears fish for salmon and scavenge seal and walrus carcasses. Citing glacial retreat and pollen, the report concluded, "By about 16,000 years ago, the North Pacific Coast offered a linear migration route, essentially unobstructed and entirely at sea level, from northeast Asia into the Americas."

Sites along the coast from deep in the Ice Age would be underwater now and mostly lost. Still, people could have been there. The textbook picture of bold mammoth hunters chasing game across a land bridge and through or over the ice is changing; now it looks like skin boat travelers. Ultimately, people would have crossed the land bridge anyway and probably attempted the ice sheets, hopping along nunataks and ice-free corridors. Yet the coast remains the most viable route into Ice Age America.

Paddling past coves and gray bedrock shoals, we were a fleet of six tandem kayaks, brightly colored, glowing under a low gray sky. We'd been dropped out here by a boat three hours out of Whittier, Alaska.

I paused and set my paddle across the deck. My older son and I sat in sprayskirts in the two cockpits of my double-seater. He was ten; my youngest, six, paddled diligently in his mother's boat a few hundred yards out from us. Others were within shouting distance. We had five children altogether, ages four to almost twelve, and seven adults, most of us parents, and one younger, childless couple. This was our tribe.

I had wanted to get closer to the ice—better representing the late Pleistocene—but the moms said no. I said we could camp several miles west of here, where glaciers calved directly into the water. The air temperature was almost 20 degrees colder from its proximity to ice—it was frigid and epic, we would have to bundle up, but it was worthwhile, I argued. We'd be kayaking among half-stable ice masses and bobbing bergie bits. But the moms decided on a constellation of islands sheltered behind a bigger island with glaciers, protected from weather coming in off the Gulf. The only ice was the occasional stranded berg, or a swan-shaped ice sculpture drifting by. The late poet and orca scientist Eva Saulitis liked to shelter her research vessel here during storms. She'd later tell me that we picked one of the most placid parts of the entire sound, a cove that she called "the Sanctuary." The moms must have known this somehow. Saulitis told me that on a quiet night, she heard the sounds of children and adults, voices and movement, even though she knew no one was there. Later, she learned that the cove had been used by Tlingit families during big hunts. She believed she'd been hearing ghosts.

Instead of paddling, my older boy skimmed the water with the blade of his paddle. That's a thing with ten-year-olds; getting them to paddle can be difficult. If he were a Paleolithic kid, I thought, he'd get smacked upside the head. But I was a twenty-first-century dad; I wondered what my son imagined in the shadows below.

A humpback whale spouted several hundred feet to our left.

"Jasper," I said. "Jasper!"

He looked up and rested his paddle on the deck as the whale's back crested slick and blue, then tipped down into its dive, the spray from its blow landing across the water like rain. The boy didn't say anything, his eyes taking it in, brain folding this away into memory. He'd seen whales before, but never so close. When the fluke lifted and slipped under, he tipped his paddle to the water and returned to his ripples.

Imagine the shock for a Beringian from Siberia roaming south to the land bridge coast and seeing an animal six times the heft of a woolly mammoth rise from the water with a misty exhale. Amazed parents must have pointed then, too. In a proto-American, Indo-Eurasian language, they would have told their children *look, look*, whether the child was looking, or not.

A second whale spouted closer. Spray went forty feet into the air and curled into mist. We heard a great, windy inhalation as it turned into its dive. The slick green-obsidian spine slid into the water. Last to go were the flukes flipping upward, casting off a cascade of seawater before slipping under.

On land, the kids never went anywhere alone, not even to pee back in the thickets of alder and spruce. Always in pairs, they wore whistles around their necks. There were bears. Scouting a nearby island for camp, we found a fresh kill in a maze of grizzly tracks—a fawn about the size of the youngest in our group, its bones splintered and head twisted backwards, eyes glassy. The bear had covered up the carcass with a light scraping of sticks and leaves.

We had a single sidearm, a .357 handgun, not the most effective for laying down a bear. It was for noise mostly, warning shots if necessary. We left kayaks at camp one day and walked into the interior of our island, lunches in daypacks, kids ringing bells and clapping, adults with bear spray, air horns. We climbed to a high point, pushing kids ahead, helping them over fallen trees, steering around spiny raspberry vines. We cleared the top with a view of surrounding islands and the white cap of an icefield pouring down through peaks, touching water.

That night, quiet fell over camp. The long day's ruckus calmed, water in the cove was still and dark, the rocky shoreline hugged by alders and shaggy black spruce. The last light went out in the last tent on a small island across from the kitchen. Stories were told, cheeks kissed. In the twilight of a southern Alaskan summer, the rest of us stayed up, tidying. As if going through the house after bedtime, we put away toothbrushes and damp socks. But we weren't in a house; any food not in hard-plastic bear boxes was packed into bags that we hung up as high and as intricately as we could, away from the bears.

Becky Ela, one of the moms, sat on a gear box, whittling a stick with a buck knife, enjoying the evening's peace. She said, "I love having a small contained existence in the wild. I lie in the tent at night with my family all around and think, I love this."

She smiled at me and said, "No offense, but probably more than you do."

Gender had been an issue from day one, our tasks divided by male and female. This was never discussed, it just happened. "We're all working, and it's equal to keeping things going, but different," Becky said. She took a breath, slivered off another curl, and told me the women did the meal planning and most of the cooking, while the men built fires, caught fish, et cetera. Rain began to fall in the dusk light as she curled back wood with her knife blade. She told me that she and my wife—the two mothers—knew where all the kids were at any moment, and knew what they were doing, who was cold, who had spilled hot chocolate all over their pants, who couldn't find their socks. It didn't matter whose kids they were. "I think that's a pretty gender-identified awareness," she said.

She seemed content, sitting with a friend out of the rain on a wilderness island. Our roles, the weather, didn't seem to trouble her.

"I think if we stayed here longer, our gender roles would evolve," Becky said. "We've got three strong, competent women who are probably just as happy fishing and paddling and lighting fires as, say, boiling Tasty Bites. Given the duration of our stay, I believe that we've just settled into roles which we know and others have come to expect— perhaps not what we would choose but what we fall into."

"You think we were the same in the Ice Age?" I asked.

Another flick of wood.

"Yes."

Nicole Waguespack, a paleoarchaeologist at the University of Wyoming, sees women's work in the Ice Age as invisible, at least as far as artifacts are concerned. Women likely used organic raw materials, plant products, leather, and sinew, things that don't survive many thousands of years. Archaeologists look for stone, bone, antler, and wood—stereotypically male items. Waguespack calls the phenomena "The Incredible Shrinking Prehistoric Woman." She writes: "Equating women solely with plant gathering is reducing their role in prehistoric societies to activities for which they may have spent little time and effort. The 'shrinking' phenomenon may not be entirely the effect of preservational bias but the inherent bias of archaeologists limiting female labor to the plant realm."

Women could have been out killing mammoths while men cradled children. The assumption that it must have happened the other way is largely a feature of modern patriarchal thinking.

When I told Becky about the incredible shrinking Paleolithic woman and noted that women are believed to leave only artifacts that vanish, she laughed. "We don't leave artifacts? Come on, we leave the only real artifact there is."

When sea levels dropped, it was time to move. The world was suddenly open, islands and continents connected by land bridges. Asia and North America formed a single, unbroken coastline twenty-two thousand miles long from Sumatra around the northern horseshoe of the Pacific Rim, connected by the land bridge and down to the bottom of South America. The Bering land bridge was the linchpin, and though you might first think of its grassy, mammoth-studded interior, it had a coast as well. This is the most likely way people arrived in enough numbers to make an archaeological dent in America.

Evidence of boat use along the Asian coast goes back at least forty-eight thousand years, when humans reached Australia. That would have required open-ocean crossings capable of carrying families, enough to form a genetic founding population. Late Pleistocene shell middens on the island of New Ireland, off the coast of New Guinea, are forty thousand years old. So are the stone artifacts on the nearby Solomon Islands. Too far to swim means boats.

Obsidian artifacts thirty-five thousand years old have been found on Kozushima Island off eastern Honshu in Japan, while a thirty-five-thousand-year-old human skeleton was found on Okinawa off the coast of Taiwan, another stretch no one could have swum.

Farther north, people may have switched from straw and wood boats to skin and wood, encountering Bergmann's Rule: The colder, the larger. Their umiat would have appeared in the Kuril Islands on the eastern edge of Russia.

I spoke with Jenya Anichenko, a coastal prehistoric boat archaeologist born in Russia and recently moved to the U.S. Anichenko, currently a researcher at the Anchorage Museum, has identified wooden pieces of a thousand-year-old umiaq, the oldest boat remains preserved in the circumpolar North. When we talked, she was living and working in Alaska. Anichenko asked if I'd ever lived anywhere completely unfamiliar, actually lived there, not just visited, but put down new roots. I said no, I was a homebody. Though I'd traveled extensively, I'd lived only in Colorado, Arizona, and New Mexico, each state pressed against the other. This constitutes *tethered nomadism*: ranging far but always coming home. Anichenko, on the other hand, had joined the ranks of true nomads, people who never return. She'd go back, of course, and visit places and loved ones in Russia, but she'd never live there again. She'd jumped continents. She said I should try it sometime, that it would greatly improve my understanding of how humans leave home, what a deep-seated challenge it is to keep going across the globe. Holding me fast in her blue-eyed gaze, Anichenko said, "Immigration is a disease."

Anichenko said I would never find the boats I was looking for. She doubts that perishable materials such as skin and wood could last ten thousand years or more. With sea levels four hundred feet higher than their Ice Age lowstand, most coastal sites are inundated.

The first people's trail along the Asian rim is marked by other artifacts, not boats. Core and blade technology, bifacial points and stone scrapers from the late Pleistocene, came up in the Kuril Islands to Kamchatka. From there, it was two thousand miles of coast and island hopping before landfall in Alaska, where Diane Hanson has been waiting for the first people with her excavation crews in the Aleutians, a catcher's mitt ready for their arrival. But how many came, and would they even be visible? Genomic modeling suggests a founding American population of five hundred people, barely enough to take hold. So few people would easily slip through a scientist's fingers.

Jon Turk kayaked three thousand miles from the Kurils to St. Lawrence Island in the Bering Strait. He took six months in a series of legs, exploring shorelines, hugging coves, and returning to useful campsites. Turk is a chemistry PhD turned science writer and hardscrabble adventurer, now writing books on long-distance exploration crossing landscapes of ice and water few would ever consider. He started this journey in a trimaran and switched to a kayak to make the rougher crossing to St. Lawrence Island. The route, he believes, follows the travels of an early people known as the *Jomon*, an ancient

Japanese culture going back at least to the end of the Ice Age. Some of their artifacts can be found in the Metropolitan Museum of Art, elegant ivory needles, hooks, and harpoons. They may have been descendants of those who first paddled to the Americas. Turk thinks the earliest Jomon must have known of this side of the world and traveled here frequently, their transient encampments now hidden under hundreds of feet of water, the Ice Age nearly forgotten.

Traveling such distances as an adventurer is one thing. Doing it with families is a wholly different proposition.

Day five, we were running low on food. The moms were the first to notice. Dads did the food-buy in Anchorage, and we'd gone light on snacks, which cut into the staple supplies. We were eating rehydrated hummus and cheese too fast. If we were real Alaskans, we'd be eating salmon, but we didn't know these waters. We dropped a crab pot, checking it a few times a day, mostly catching starfish. Over the next few days, we had to ration and dig deeper into wild foods. Steamed clams for breakfast, lingcod caught on a line and sizzled in a pan for dinner. We had a biologist on the trip, Irvin Fernandez, who specializes in eating from the land in unfamiliar places—spiders in Cameroon, and tree-beetle grubs in Central America. He was just the man a tribe like ours needed.

I'd traveled with Irvin for decades, eating bugs and seeds in the desert with him. We killed a jackrabbit in Arizona with a slingshot, gutted, skinned, cooked it on a *palo verde* fire. Traveling the Mexican coast of Sonora by foot, we ate almost every living thing we could find along those parched desert beaches, emptying slugs and silver-dollar crabs onto a tarp to decide what was viable and what wasn't. Irvin had an admirable way of experimenting with his mouth, taking in unfamiliar foods, usually faunal, and chewing them to pieces with a quizzical expression on his face, asking questions of every taste and texture. He believes it is a coastal way of eating, something that comes naturally to families of the Pacific Rim. It is also something you learn while foraging in unfamiliar foodscapes. What he didn't want he spat out: shells, bones, exoskeletons, acidic excretory sacs, the head-caps of grubs.

The kids watched, captivated. It was like science class, studying antenna sprouts and knobby eyeballs of prawns we caught in the pot as Irvin showed them how to eat heads whole. He said yes to one thing, no to another. *Yes* to the many mollusks we dug up in our low-tide clamming industry. *No* to a twisting, two-foot-long annelid worm that a farm kid from western Colorado wanted to put down his sister's shirt.

The younger kids, including Irvin's four-year-old son, proffered seaweed to the group, collecting different species based on the biologist's instructions. They toasted *Palmaria hecatensis*, a red-ribbon seaweed, with olive oil and salt. They called it *sea bacon* to entice us. We ate until we couldn't take anymore.

I believe that's why coastal travel would have always been preferable to the interior: free food. Use of marine resources was common in the Ice Age. Isotopic measurements of bone collagen from twelve-thousand-year-old human remains in a cave in North Wales, United Kingdom, indicate that 30 percent of their dietary protein came from marine sources. Technically, Ice Age people weren't hunters or gatherers. They were like us in our kayaks: foragers, searching widely for food and provisions. In the interior it was eat or be eaten, while here there was food in the water and on land. It was hard to go hungry. You could dig in the wet tidal zone for mollusks, gather seaweed, fish, prawns, berries, deer, moose. At Ice Age coastal sites in California, Peru, and Chile, people ate mastodon, salmon, seaweed, deer, and sea lion. They weren't maritime specialists, unlike most coastal dwellers from the last several thousand years. They inhaled the coast as they went, whatever land or sea offered. Every degree down the face of the globe brought something new.

Children, we discovered, were a great bear deterrent. Adults scouting other islands found fresh bear sign all over the place, tracks and scat everywhere. With kids, however, our camp felt off-limits, the closest fresh bear scat about half a mile inland. Once we took over a cove with tarps and squealing, running little monsters, the bears cleared out. I pictured them huddled at the opposite end of the island, waiting for all the yelling and stomping to cease.

I studied each of our children, wondering who they'd be in a tribe on the far side of the world. Any tribe would need hunters and heal-ers, people who knew the plants, what was edible and what wasn't, the bark of mountain alder good for arthritis, and willow for headaches. How to eat, and how not to get eaten. Someone would have to have an eye for sickly mammoths in the hunt, another would need to be good at starting fires with a hand drill, catching kindling under a crisscross of greasy megafauna bones on a wild Pleistocene coast.

What would my own boys have done in this ancient tribe? They stuck their noses into bear prints, ate leaves, and held up driftwood, aligning themselves with horizons. In a half-hearted rain, my younger boy, Jado, was down by the water talking to animals. Six years old, he stood at the edge of the tide in black rubber boots, holding a barnacle-encrusted rock. He cupped it near his mouth in both hands and hummed to it.

I approached the boy like a modern father; I flipped on my camera and asked what he was doing.

Jado looked up, distracted, dots of mist caught in his eyelashes. Glancing at the camera, he said, "I'm humming to make them come out."

"To make what come out?"

"The snails."

He tipped the rock toward me, showing a couple small, black periwinkles.

"So you hum, and the snails come out of their shells?"

"Yeah, but it takes an hour to get them to come out."

He turned and scurried away.

He Who Talks to Snails. You would have needed one of these in the Ice Age at the edge of the known world. The spirit of an animal can kill you. It is not human. It is not concerned with our enterprises, so we must be concerned with its. When you are pure in wild places, you understand what is not just us, what is beyond our self-admiring sphere. The first people would have hedged every bet. Somebody needed to talk to the animals, calling them in for the hunt, fending off the most fearsome and dangerous with a prayer, an offering, a word of thanks to land and sea. Whatever it took. Shamans were needed.

The earliest known shaman comes from twelve thousand years ago, well within the brackets of the Ice Age. The remains are from a late-Pleistocene excavation in the Galilee region of Israel. A roughly forty-five-year-old female was buried with fifty complete tortoise shells, the leg of a wild boar, an eagle wing, cow tail, a leopard, two marten skulls, a complete horn core from a male gazelle, and a single human foot. The unusual accompaniment of parts suggests to researchers that she held a position of spiritual import between worlds.

In North America, the oldest suggestion of shamanism—not counting the simple presence of ochre—comes from the last few thousand years: ritualized twig figurines of animals in the caves of the Grand Canyon, or wildly pecked and painted canyon walls in Utah and California, people turning into birds and animals, liminal creatures between one world and the next. You see faces coming out of the caves of Jamaica, their rounded features like ghosts not quite in the light, but not in darkness, as if peeking out at the created world.

Shamans are the ones who went into the shadows and did whatever could be done to ensure the survival of the people. You would have needed them to get you through.

With its gentle breakwaters and shelters, the Ice Age coast of Alaska would have been a slightly colder version of what is here today. Based on sea-floor topography, with sea levels four hundred feet lower, the coast along the land bridge was edged with hundreds of small islands. Kelp forests wove through the waters, forming the northernmost stretch of the "Kelp Highway," a ribbon of high biotic potential that stretched from Japan to Baja. An unbroken stretch of luxury: an invitation to North America from Asia.

Alaskan kelp beds in the Ice Age may have been more productive in the cooler waters, their ecology richer than today. Now-extinct sea cows, manatee-like mammals thirty feet long and weighing several tons, would have drifted near the surface. Even with an ice-locked coast, this would have been far easier foraging than the heart of a five-million-square-mile ice sheet, or some crumbling, glacier-scoured corridor from Yukon to Montana.

In the morning, shucking clams into a bucket in the rain under a tarp, conversation arose. What if we were the only ones out here? This was the farthest edge of the human species, no one ahead of us, and no one close behind. We'd have to decide who got extra food. The childless man who helped out the parents would have to take precedence, we agreed, his reproductive vigor greater than the rest of the dads. He was a muscular street cop from Denver, and as we talked about this, he shot a nervous glance at his tall girlfriend, a firefighter in training. Neither of them were ready to have a child, or even to enter into a conversation like this.

The orchard farmer pointed out that he has to schedule tree planting ahead of time to keep his business healthy. He plants six years in advance of a harvest, and the same would probably be necessary for our group. At least every six years, a new child would have to arrive.

My wife, forty-six-year-old mother of two, looked at the tall girlfriend, childless and thirty years old, and said to her, "I believe you're up."

The woman replied, "Why don't we use you up first? You look like you still have another one in you."

We were off to a good start. With a generic European blend and a first-generation American-Filipino dad, a half-Korean mom, and a childless African American man, we had a strong mix, the kind of genetic foundation you'd hope for when starting a new world.

Regardless of our ethnic, genetic vitality, we'd be on the path to inbreeding within two generations. Survival rates among newborns fall sharply after direct bloodlines begin mating with each other. Baby girls would be less likely to survive. Female physiology takes more energy to maintain; males, cheaper to produce, would be sent out like the last sparks, a burst of sperm donors looking for females as we drifted toward extinction.

The magic, minimal number for human colonization is one hundred and sixty. Twelve is far too few. Forty is possible, with a strict mating regimen, but it still skirts dangerously close to extinction. A 2014 study in the journal of the *International Academy of Astronautics* looked at a five-year colonization mission into space, estimating that between fourteen thousand and forty thousand people would be the safest guarantee for long-term survival. Those numbers would maintain good health over five generations, enough to last through at least one severe population catastrophe before reaching a location where they could spread out.

In a group of twelve, we'd need to find more people within one or two generations, or we'd be lost. Monogamous marriages would be disbanded, social norms broken for the sake of a richer gene pool.

That morning, of course, we were speaking hypothetically. Who knew that within a year, two of the married partners in our coastal tribe would be drawn to each other, and I would become a single dad? At the edge of the world, tribes come apart; survival takes its toll.

Down the coast, the scrawny black spruce woods of the North turned into temperate rainforests towering with hemlock and cedar. The world felt as if it were opening, blossoming. Away from icebound wastelands and Arctic coasts, the South has always had more of everything good.

A Boeing 737 half-cargo plane making a milk run through the unroaded towns of southeast Alaska dropped us in Wrangell, a coastal town six hundred miles down-coast from Prince William Sound. Three parents and four kids had left us by now; they went back to their lives after we'd been motored off the Sound from a hired boat, three wave-smacking hours back to the glacial enclave of Whittier, Alaska. Our tribe had shrunk to two boys, my wife, and a support crew of two friends.

Our umiaq was now a rented minivan with an interior stripped to metal and a bench seat in the back. It had old, brown blood on the seats: elk, deer, or bear. The back was dirtied with needles and bark from hauling wood. We drove it out of town and camped in a forested park along the beach where we readied food for an inland backpack. A stream flowed through rocks and giant red cedars. Our tents had popped up in the woods like mushrooms, and in the morning we loaded packs, getting ready to leave. My wife, Regan, stuffed sleeping bags while I cooked breakfast on the camp stove. The boys had made a kingdom at the top of the twenty-foot-tall root bundle from a fallen cedar, its trunk the size of a bus. This is what kids do when they land in new territory. They make it home.

I heard a cry and saw Regan sprinting toward them. Only one boy was up in the roots, the other was gone. I flipped the stove to off and ran.

Regan knelt in the stream below the fallen tree. Our youngest, the Snail Whisperer, lay facedown in the water. She slid an arm beneath Jado and cupped the back of his head as she rolled him over. His eyes fluttered open, head pumping blood. Stream water, gravel, mud, and cedar needles matted to flaps of skin, a door the size of a cuckoo clock's open in the middle of his forehead. His skull was exposed.

You worry about your children all the time—the tone of your voice, the chemicals in food, the socks they can never seem to find. You build a nest for one purpose, and you shall not fail.

But now, at this moment, I did not worry. My body was only a tool. I sprinted for my pack, calling to our friends, the firefighter and the street cop: "Head injury!"

I yanked open my pack, hands diving in. The backcountry kit was buried too deeply; we needed compresses, *now*. A stack of boxer shorts came out in my fist, washed yesterday at the coin-op in Wrangell. These would do. I ran to the water where Regan was still looking down on Jado, telling him, *You're OK, you just hit your head really hard.* I could tell from behind that she was smiling at him, his eyes locked on hers. I laid the compresses above his eyes, folding the doors of his forehead closed as she checked toes and fingers, and Jado answered calmly, shocked.

When we pulled into the clinic parking lot in Wrangell, it was clear and sunny, the kind of day you pray for in Southeast Alaska. Regan held the boy, his head packed in bloody laundry, as one nurse calmly got on the phone and the other, equally unstirred, pulled out trays and readied the gurney. Jado was fully conscious, no tears yet, and as Regan laid him down, the nurse pulled off our compresses around dark red clots, exposing what looked like a volcano in his forehead.

The doctor arrived within half an hour wearing a denim vest, cowboy boots, and a battered leather multi-tool on his belt. It was Sunday morning. He was unshaven. The only two nurses in the clinic were swabbing and patting to slow the bleeding.

"Let's have a look at you," the doctor said.

Looking upside down at the laceration in the boy's forehead, pulling on latex gloves, he introduced himself and said with a smile, "Did your brother push you?"

Jasper's eyes widened with shock. His mother touched his hand, smiling, *He didn't mean it.*

Jado had been too stunned and bloody to cry. Then a nurse handed the doctor a syringe with a needle, and he began to resist. With his head on a pillow, he started asking rapid-fire questions, anything to keep that needle away.

The nurse held down one shoulder and the second nurse came in to hold down the other. Soon both Regan and I were pushing him back down as the doctor moved in with the needle and the boy began to wail, rattling every glass in the room. He was going to be fine.

It has never been easy to lose a child, not now, and not in the Ice Age. I'd thought about this on the kayak portion of the route, finding the fawn crumpled up from a fresh grizzly kill. The girl with limpets on her fingers. The boy talking to snails. When these are the ones you take care of, the space between life and death can be uncomfortably thin. Any of us would have died for the kids in our group, regardless of whose kids they were. A tribe is technically defined as a social division consisting of interlinked families working toward a common goal. In a simpler sense, a tribe may be the people whose children you would die for without hesitation.

Human remains throughout the Paleolithic tend to come from children, not adults. Consider the oldest three people found in connection with the first American colonization: a Siberian discovery, known as the Mal'ta Boy, 24,000 years old; the approximately 13,000-year-old Anzick Boy excavated from a cave in eastern Montana; and the 12,900-year-old remains of a young teenage girl who was recently found in the bottom of an underwater *cenote* in the Yucatan.

The boys from Siberia and Montana were toddlers, both carefully buried, both with red ochre. The girl found in Mexico was older, and she may have gotten lost or fallen, and died alone. She represents the southernmost human remains from the Ice Age people who started in Eurasia.

In my mind, these three kids seem like a myth: magical siblings spanning eleven thousand years, two brothers and their older sister from Beringia coming into a new world. In a watery cenote, the girl returned to the earth, sewing into the land the arrival of the first people.

Anzick Boy was between one and two years old. Discovered in 1968 during construction on the Anzick family's land near Wilsall, Montana, his skull and bone fragments were in a horizon of rock flakes and large, artful projectile points from what is known as Clovis tech-

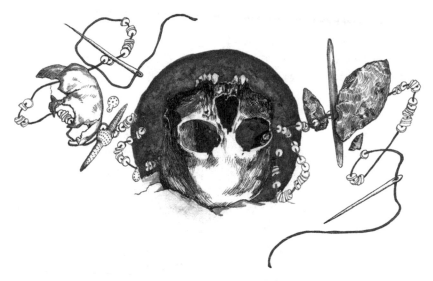

Adornments and a skull bone of the Mal'ta Boy from Siberia, the complete skull of Naia as it was found at the bottom of *Hoya Negra* in the Yucatan, and stone points and a bone rod found in association with the Anzick Boy in Montana

nology, bifaces covered in red ochre and laced across his bones. The bones were roughly treated after discovery, kept in boxes and handled by many. Their radiocarbon dates were contaminated, but a piece of antler buried above the boy is about thirteen thousand years old, putting him later in a litany of arrivals. Ninety-eight percent of the DNA extracted from his remains belonged to bacteria that invaded the bones during decomposition. Two percent was human, though, enough to construct the oldest recorded human genome from the Americas, one more closely related to Native Americans than to any other group worldwide. It is also related to gene flow from the boy interred twenty-four thousand years ago at the Mal'ta site, a complex hunting village in eastern Siberia.

The Mal'ta Boy from Siberia wore a bone-bead necklace with a "bird-shaped" pendant, and around him were burial artifacts including a Venus figurine, a female form fashioned in a style known from Siberia to France. This boy was part of a movement across the Mammoth Steppe, similar artifacts connected over thousands of miles. His genes belong to a now-extinct population that contributed to Siberians, Native Americans, and the Bronze Age Yamnaya people, known also as the Ochre Grave Culture, who lived in Eastern Europe north of the Caucasus between the Black Sea and the Caspian Sea. His DNA carries no signatures from Central Asia or East Asia, which means he was pure Beringian. Only later did coastal genes reach his descendants on their way into the New World.

The Yucatan girl represents the oldest, most complete human remains in the Americas, a mostly intact skeleton discovered disarticulated on the bottom of a cenote. *Naia*, they called her, Greek for a water nymph. Her skull, resting upside-down, was found by divers at the bottom of a spring called *Hoya Negra*, the black hole. She may have become lost in the cave and fallen into a chamber more than a hundred feet below. When sea levels rose, groundwater was pushed upward, drowning the chamber, encasing Naia in fresh water along with the bones of a gomphothere—a kind of extinct elephant—and an array of extinct predators that must have fallen in as well. Given the watery state of her preservation, only mitochondrial DNA could be extracted, the matrilineal bloodline representing yet another Eurasian and Native American connection. Part Mal'ta Boy and part Pacific Rim, she belongs to a thread of mothers who came from the other side of the world.

Paleolithic human remains tend to be children, not because their bones are somehow more long-lasting, but because their bodies were more likely to be cared for. The Mal'ta and Anzick boys were intentionally buried, rather than left to the elements. Naia was preserved by chance at the bottom of her hole.

The doctor closed Jado's wound with sixteen stitches. Soon, the boy was sitting up with a blank look on his face, the letter "Y" ladder-tracked down the middle of his forehead.

We went back to our camp that evening at a beach outside of Wrangell. Tide logs were stacked up, whole trees washed in over the decades. We dropped in a tangle of long, naked tree trunks, sheltered from an evening breeze coming off the water. Regan sat against an ocean-polished cedar, our little boy in her arms. Jado's face was drawn from painkillers, and his older brother crouched beside him, holding his feet in their small socks.

In the bronze light of one of our last Alaskan sunsets, he was still a bit stunned. Regan brushed her hand around his belly like a slow clock. She sang softly to him.

I have wondered if the ochre, the fire, and the care of bodies represent layers of sadness, if they tell us something about love in the Ice Age. Adults may have been left on the ground after death, to be scattered by weather and scavengers. Children were different, their loss felt more acutely, cared for even in death. They were, they are, true hope.

Three children were buried at different times in the floor of a hut in central Alaska, at what is called the Upward Sun site. The hut was made of poles and skins eleven thousand, five hundred years ago. The next oldest after the Mal'ta-Anzick-Yucatan trio, these three lived and died in the Tanana River Valley, the next major drainage south of the Yukon. Two are infant Ice Age girls, who came from two different mothers and were both buried with grave goods.

In a layer above the two baby girls were the cremated remains of a child about three years old and of unknown gender. Two pieces of red ochre were found among the burned bones. Based on the soil horizon, he or she must have died after the two infants, possibly in a different season. The cremation fire was set in the hut's central fire pit, atop layers of earlier fires, salmon bones, muskox, horse. It was the last fire lit there; after this, no one ever came back. The cremation happened at the warmer end of the Ice Age, when Alaskans could burn wood instead of bone, and so the fire was made of poplar wood. It lasted from one to three hours and was never stirred; the skeleton remained in position, intact. I imagine the adults leaving through a door flap, letting the light in, shafting through the smoke.

The doctor with the stubble and cowboy boots told us we could backpack, to treat the boy as if he were healthy, just make sure he didn't knock his bandaged head on anything. On a point of land a thousand feet above inlets and islands, we set up camp, looking down the ragged chain of the Alexander Archipelago. Waters were calm, streaked now and then with the courses of gill-netters circling for salmon. We saw very few boats or people. The land before us lay empty. Beyond blue courses of water, beyond the horizons of forested islands, stretched the rest of the continent, south of the ice sheet.

Eight hundred miles left to go.

We traveled the rest of the way on the *M/V Columbia*, the 418-foot-long flagship for the Alaska ferry system, room for 280 cars in its cavernous two-story belly. It was named after the nearby Columbia Glacier, which was named for Columbia University and its early expeditions mapping ice in southern Alaska. The university itself was named in 1784 in honor of "Columbia," a quasi-mythical female symbol that came to represent America, the New World, and eventually the political fervor of Manifest Destiny. In an 1872 painting, she appears as a woman drifting over the prairie in a white flowing dress, a spool of telegraph wire over her arm, leading miners, railroads, stagecoaches, and homesteaders westward while bison and Indians flee before her in terror. Columbia was, in turn, named after Christopher Columbus. The name on the side of our ship bore more than a whiff of colonization, the conquest of and expansion over land and whatever people and animals once inhabited it. It is a concept this side of the world has known for a very long time, that urge to fill an emptiness that never existed, except in the minds of those yearning for untamed vastness.

Our giant umiaq rumbled through the Alexander Archipelago of Southeast Alaska, where glaciers pour between islands and peninsulas offering shelter, breakwaters from the wild Pacific. At night, the twin 9,000-horsepower engines thrummed us to sleep in a tent on an aft deck, the metal under our sleeping bags beating like endless galley drums along the Ice Age coastal route.

Regan, prone to seasickness in even calm waters, had flown ahead. Our tribe was down to three, a father and two sons, one still bandaged from an injury. That might be as good as you could have hoped for, after a long journey down a ragged coast in the Pleistocene.

I found myself leaning on the wet railing, watching the darkening, temperate forest glide by, defiles of fjords and islands on either side, waterfalls cascading into the sea. How many eyes looked back, I wondered. Grizzlies, black bears, caribou, moose, wolves, elk. Orcas sank away unseen, and humpback whales breached, sending up jets of mist.

In the last seventeen thousand years, this coast has only grown richer, denser, more green where glaciers have rolled back and back. Having spent most of my life in drier, sunnier places, I'd fallen in love with this coast and its density of life. It drew me on, as it would have drawn tundra hunters. Northern people are known for their mechanical ingenuity, repairing machine parts with bone or leather, calculating the friction-wear on skins used to replace belts on engines. In the Ice Age, they would have easily whipped up boats, lashing skins and wood together to reach some richer-looking island or bay that could only be accessed by water.

Prince of Wales Island, a hundred times the size of Manhattan, passed on the starboard side of the ferry. The island, made mostly of karst limestone, has caves where spelunkers and researchers have found Ice Age bones, one of the few places along the Northwest Coast not entirely swallowed by ice during the last glacial maximum. Animals lived here back then, most of the caves revealing an abundance of Pleistocene bears. A report from one cave reads, "Along with four black bears, the remains of at least three brown bears were recovered, one of which was gigantic." The caves also produced late Pleistocene fleas, nematodes, and signs of black spruce, marking this corner of Southeast Alaska as an Ice Age refuge, a place where people could have landed by sea even during a glacial maximum.

Prince of Wales Island was one of the genetic seeds that would repopulate coastlines as ice retreated. While fossils on nearby islands experienced a hiatus during the last glacial maximum, they kept going on Prince of Wales. Along with giant bears, the remains of Ice Age caribou were found in caves as well as bones of ringed seals, indicating the presence of pack ice.

One of the island's caves, called On Your Knees, held the remains of a man who was about twenty years old when he died 10,300 years ago. The oldest known human remains found on the Northwest Coast, long after the first people would have come through, ended up in the back of the cave. Large bite marks on the fresh bone suggest that his death might have been violent, or at least that his body was scavenged and dragged back here.

The cave's entrance has produced thousands of flakes and stone tools, including stemmed projectiles and leaf-shaped blades, their dates less certain, but likely in or near the Ice Age. Some of the tool-stone found on the island suggests that people were living here by eleven thousand years ago, traveling by boat and land over a hundred miles to particular rock sources and bringing back regular supplies. They must have known the country at least well enough to be aware of the best rocks within a hundred miles, some of the sources far inland between glaciers.

The human bones from On Your Knees were claimed by the local Klawock and Craig tribal governments, of the Tlingit Tribe. They named the person *Shuká Kaa*, in English, "The Man Ahead of Us."

Tlingit is an Indigenous coastal group of hunters and gatherers. They erect totem poles carved with ovoid eyes of animals, their art similar to that of the indigenous Ainu from the northernmost island of Japan on the opposite side of the Pacific. Their genes connect them directly to Shuká Kaa. When the bones were found in 1996, local tribal representatives were immediately notified. The two Tlingit tribes agreed to have the remains studied, concluding that the body had not been intentionally buried but was dragged into the cave by scavengers. How the dead are treated makes a difference. This man had not yet received a proper burial.

Washington State University molecular anthropologist Brian Kemp found that Shuká Kaa's closest genetic relatives can now be found along the West Coast from here to the bottom of Chile, defining an ancient route of migration, one that was probably established long before The Man Ahead of Us lived and died.

Shuká Kaa was buried in 2008, his bones and teeth placed inside an intricately carved and painted wooden box, the red image of the man on the front of the box guarded by Raven and Eagle. Prayers and smoke were given to the remains, and the box was placed in the ground at an undisclosed location. A gray slab of rock was erected at the site, and on it these words were incised:

SHUKÁ KAA
WE HAVE LIVED IN SOUTHEAST ALASKA
SINCE TIME IMMEMORIAL
SHUKÁ KAA IS TESTIMONY TO OUR
ANCIENT OCCUPANCY ON THIS LAND.
BORN CIRCA: 10,300 YEARS AGO
DIED CIRCA: 10,280 YEARS AGO
SEPTEMBER 25, 2008

Ernestine Hayes is a Tlingit memoirist, scholar, and state writer laureate for Alaska. She spoke with me about these bones, and the people who came before her. She said, "Raven found people in the clamshell, and people who are here have always been here since time immemorial."

I asked what these words *immemorial* and *always* mean, and she said they indicate an actual bracket of time, something that both story and science can help define. She said that this was not some impossibly mythic past, but something the people remember, not necessarily a land bridge, but another kind of emergence, a shell washed up carrying the first people.

"The land bridge is what people who follow the religion of science tell themselves," Hayes said. "I think both can be true at the same time. One just happens to be dominant, a product of colonialism."

Hayes mentioned a well-known Northwest tale, how the raven used to be white, and came to be black. Like all good stories, it means more than it seems, and she thought it might hold answers for my questions about Ice Age arrivals.

Raven, the trickster god, lived along this coast. Raven was white then. Not that it mattered; this was before the world had light. All was bathed in a skeletal darkness where you could make out the softly glowing, bluish shadows of trees and figures, but little else. There were no days, no nights, no seasons, no stars. At least not until Raven came along.

A figure named Grandfather kept the sun, moon, and stars inside a wooden box in his house. The box would have looked something like the one that now holds the remains of Shuká Kaa, painted and decorated. In some stories, it is said to have been carefully inlaid with small cowry shells. Raven tricked his way into the house through a classically complicated and devious series of events involving Raven impregnating himself into a young woman, who gave birth to him inside the house. Grandfather believed he'd been given a grandson—a spoiled and demanding grandson. When Raven was old enough and saw his chance, he threw the lid off the box, and light flooded out. Raven was brilliant in the light, as bright as snow. He snatched the sun, moon, and stars and tried to fly out through the smoke hole. But he got stuck in the hole, caught by what he'd stolen. Grandfather rushed in and found the thief trying to get away. He built a fire beneath the struggling bird. Finally, Raven burst through the hole, trailing sparks and smoke behind him, and released the celestial bodies into the sky. Singed and burned, Raven was now black, shiny as charcoal, the color we see today. But the sky that we know was now in place, too.

Hayes told me this is a story about the end of the Ice Age. "We know that animals that live around snow and ice are white in color for much if not all of the year," she said. "Along with the retreating of ice, the white bird went away with it, and black raven came up to explore newly uncovered ground."

I asked Hayes if anyone was here before the Tlingit. Archaeologically speaking, I said, there should have been people here before Shuká Kaa died, 10,300 years ago. I was looking for people who were on this coast when Raven was still white, when the world was being made.

She said, "The oldest stories are about people who came from inland when the ice melted. They found people who were already here. Those people probably came by water."

These were the people I was looking for. The accumulation of evidence suggests that the first migrants into the Americas came along multiple routes and at different times before, during, and after the height of the Ice Age. It was not a single, small colonizing population, but many separate arrivals. When Hayes mentioned another people here when the Tlingit arrived, I used the words *first Americans*. She reminded me that this was not America then. She implored that I think differently about the history of this land, saying that people were not *becoming American,* but *becoming Tlingit, becoming Navajo, becoming Lakota*. Some lived in the thick of the Ice Age and some arrived to see it end.

In college, I spent a summer on a $700 grant along the Northwest Coast. It was my first time in the Pacific Northwest. I was nearly broke and barely old enough to drink, but I had a driver's license, so I got the hell out of where I was living, as far away as I could go. Perhaps it was my D4, the dopamine receptor that marks wanderers; perhaps it was the memory of my mother's highway umiaq urging me to go. I headed north, traveling in bush planes from Squamish to Bella Coola up the British Columbia Coast, almost to Alaska, trading labor for flights. The grant was for an interview project with Canadian bush pilots. I flew in Beavers and Cessnas, taking off with an old, one-eyed pilot in Williams Lake who lit the engine on fire as we left the ground. He flipped switches until the flames sucked back in, then grinned at me, saying, "That's what happens when an internal combustion engine becomes an external combustion engine!"

I helped carry boxes, groceries, mail, medical supplies. As we flew over mountains and blue inlets, I saw more wilderness than I'd ever imagined, nameless mountains and glaciers folding over each other. It felt like the immensity would burst my heart. I saw waterfalls thin as floss coming down steep mountainsides, their clean, gray waters milking the blue-green ocean below. My suspicions about the world were confirmed: It was wild and empty.

I stayed longest near Bella Coola, a small town sixty miles up a narrow inlet. It is the seat of the Nuxalk Nation, which has fought for at least two centuries for their territory, battling first the Russian explorers and hunters who tried to claim it, then smallpox epidemics, then the arrival of highways and airplanes. The entrance to their red-painted lodge is made into a mouth at the base of a totem pole. My memory may be faulty, but I recall a storytelling evening, looking up into the rafters through the drifting smoke and seeing a small, mirrored disco ball, left over from the community dance.

Recording in notebooks, taking film pictures on a tank-like Pentax, I spent every other day or two in the air, mapping the land with my eyes, taxiing into logging camps and Native villages, landing on water, snow, or washboard dirt strips. When I wasn't flying, I sat at my camp beyond Bella Coola watching bald eagles bicker over roosts, their great, dark wings slapping at one another.

Using driftwood and pieces of discarded pier, I built a small hut above the cobbled high-tide line. It was just large enough to sit inside or curl up in for a nap. I hung interesting items around the entrance using found fishing line: a couple eagle feathers, bits of wood, pieces of colorful plastic. When it rained, which seemed to be most of the time, I sat in the hut, evening clouds rolling through, sea and sky the same blue-green. In the crooked oval mouth of my shelter, I felt like a shaman. I was the tip of the rat's whisker, finding the way, leading edge of my people.

For the Nuxalk who gathered in Bella Coola's hall, coming in and out through the mouth of a totem pole, I was simply another arrival in a land that has seen many.

The entrance to Bella Coola slowly opened and closed as the *M/V Columbia* rumbled by. I wondered if the old wharf was still up there, and I thought of what this place had looked like from the air, a maze of fjords and mountains, what is called the Great Bear Rainforest extending along two hundred and fifty miles of the coast of British Columbia. West of here, the islands are crumbs and slabs shrugging off into the Pacific.

About fifteen miles west of the inlet to Bella Coola is Triquet Island, a gob of rock and temperate rainforest, a mile from side to side. An excavation there found an Ice Age settlement under layers of peat and soil, where, based on an array of samples, a carbon-rich fire hearth was used sometime between 14,086 and 13,613 years ago. Researchers led by Alisha Gauvreau, a PhD student with the University of Victoria's anthropology department, have uncovered wooden artifacts still intact from these oldest layers, bits of wood fishhooks, a hand-drill for starting fires, the throwing arm of an atlatl, carved wooden points, and a carved stick wrapped in bark. Gauvreau presented her finds in 2017, pushing settlement dates along the coast here further back than ever before. These may be the people the Tlingit ancestors encountered when they arrived on the coast north of here around 10,300 years ago, the people that Hayes said "came by water."

Ten miles down the coast from Triquet Island is Calvert Island, where, if the date holds, the oldest human footprints in North America have been discovered. Like Triquet, Calvert has also maintained a steady sea level since the late Pleistocene. University of Victoria archaeologist Duncan McLaren of the Hakai Institute directed a crew at a beach site peeling back layers of gravel and sand. They opened a group of human footprints associated with a nearby stone-lined hearth just above the high-tide line. Charcoal from the hearth came back dated 13,200 years old, the possible age of the prints. People had stepped in a fine, clay-like mud that took their impressions, leaving twelve distinct tracks: an adult, a second slightly smaller adult, and a child. A left foot and a right foot show the gait of someone walking, the heel pads, toes, and the arch of the foot those of an Ice Age person on the B.C. coast.

The oldest human footprint in South America also lies along the coast, a child's track found at Monte Verde along the Pacific edge of southern Chile. The excavation of Monte Verde revealed a robust human component 14,500 years old, and less robust sequences going back much farther.

As the *Columbia* continued through widening waters, I was aware of the rest of the coast ahead, a route unfurling beyond our destination, one of the longest and more or less straight-lined coasts in the world stretching from here to the bottom of South America. The direction we traveled felt like a drumbeat, the rhythm of a course of migration, written in the water.

The West Coast was a relatively busy place in the late Pleistocene. Thirteen thousand years ago, on the Channel Islands off Southern California, people lived among and probably hunted horse-sized pygmy mammoths and utilized stone tools of Pacific Rim stemmed tradition. Such finds are strung down the Peruvian coast to Chile as well. Remains of deep-sea fish have been found in Peruvian excavations, evidence that people were not spearing or netting from shore, but were using some sort of craft to reach open water. In from the arid coast of northern Chile, a trenched red ochre mine was first gouged out of an iron oxide vein twelve thousand years ago.

The history of the colonization of South America is wholly different from that of the North. South America in the late Pleistocene was the farthest large, habitable place on Earth, reachable only by small and vigorous parties. Compared to the front door offered by the Bering land bridge and its coast, the Southern continent may have taken more D4 to reach. It is doubtful that Polynesians arrived then, their gene pool reaching the Americas only in the last few thousand years. These were coastal travelers, and they came down from the North.

One study of Native Americans aimed at solving the riddle of the first people found that their genetic similarity to Siberians decreases the farther they got from the Bering land bridge. This is a chromosomal trickle-down from north to south, people beating through geographic gauntlets, their genetic connection to their starting place stretching thin the farther they went.

Until three million years ago, South America was geographically separate from North America. When tectonic changes lifted the Central

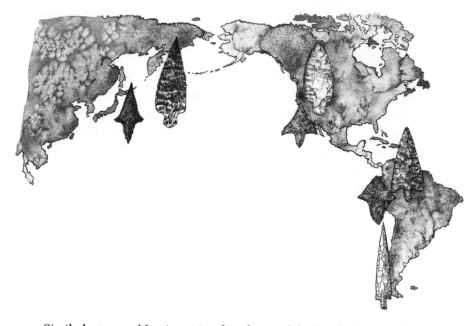

Similarly stemmed Ice Age points found around the Pacific Rim, clockwise: Incipient Jomon tanged point ~15,500–13,000 years old; stemmed points from Ushki Lake ~13,000; Buhl burial ~12,500; Channel Island Barbed ~12,000–8,400; Lower Amazon 12,500; Quebrada Jaguay ~12,400–10,500; and Paijan ~12,000–9,500.

(Adapted from "From Asia to the Americas by Boat? Paleogeography, Paleoecology, and Stemmed Points of the Northwest Pacific," Jon Erlandson and Todd Braje, *Quaternary International* 239, 2011.)

American land bridge, land animals passed between the two continents in what is known as the Great American Interchange. In its time apart, South America had bred its own pair of sabertooth cats, *Smilodon gracilis* and *S. populator*, the two species sundered by the chain of the Andes. Giant armadillos and club-tailed glyptodonts with shells of hardened hair made it as far as Mexico and the American South. A hoofed, camel-like animal lived below the equator, larger than a horse and with an extended proboscis like a short trunk, its nasal opening above the eyes on top of its skull—remembered now only in the fossil record.

With the two continents connected, late Pleistocene boat travelers had no reason to stop their southern advance. It was a long, unbroken line. The stone projectiles they left along the way were stemmed, with the nub of a tail sticking off the bottom for securing to a shaft.

Stemmed projectiles are found at On Your Knees Cave in Alaska and at Cueva Fell in southern Chile, an eleven-thousand-year-old Paleolithic rockshelter near the tip of South America. They have appeared on the coast of Peru, California, Oregon, and British Columbia, as well as across the ocean in Japan, Kamchatka, and on the Kuril Islands. This is the earliest sign of a singular people connected by a unique Ice Age technology along one contiguous coast.

Kelp ran out in the warmer waters around the southern tip of Baja California. This point of land in Mexico is relatively abundant in Paleolithic material, as if it were a diving board, launching the people southward in search of cooler waters where the kelp would resume. They would have had to paddle another three thousand miles to the top of Peru before kelp beds reappeared. Evenings would have begun to cool as people traveled south of the equator. Chile would have begun to feel like Northern California, but with different plants. Around the modern-day northern Patagonia city of Puerto Montt, glaciers emptied into the sea, blocks of floating ice populating the travelers' route as they found themselves back in the sort of world they once knew in Alaska. Perhaps they believed they were making some kind of a circle, or a spiral, the world turning back on itself.

The open-air site of Monte Verde, along the southern coast of Chile, is one of the key Ice Age human appearances in the New World. It is one of the early emergence sites: a village, or at least an encampment returned to over thousands, if not tens of thousands, of years. It lies along a creek set back from the coast. There are signs of complex and multilayered occupations, with woven fish weirs and the remains of twelve hide-covered huts, with 112 wood posts used in their construction still partly intact. These rectangular huts—floor plans ranging from 43 to 170 square feet—were joined along their sides and arranged into two parallel rows. The layout of stone and bone artifacts on their floors, hearths, trash middens, and work stations suggests the places were residential. A thirteenth hut was found nearby, the floor plan wishbone-shaped, its artifacts hinting at the preparation and use of medicinal plants: a healer's dwelling, perhaps a shaman's hut.

The foods eaten at Monte Verde ranged from seaweed to mastodon. The people were generalists, hunters, gatherers, fishers, scavengers. No bones or teeth have been found, so their DNA has not been extracted. Who they were is unknown. For now, they are just people, but their projectiles indicate a Pacific Rim connection by way of the Siberian coast.

Vanderbilt University archaeologist Tom Dillehay, who has led the multinational, interdisciplinary excavations at Monte Verde, wrote, "Assuming that other late Pleistocene people operated under similar subsistence and settlement practices, our data imply that if groups traveled along the Pacific coast, they may have migrated slowly and exploited the interior resources of the hundreds of river basins descending the long mountain chain from Alaska to Tierra del Fuego."

Migrating slowly, as Dillehay suggests, would have allowed them to learn of food resources and places to find good toolstones. But to have reached Monte Verde ahead of most other sites in the Americas, they had to have been driven, pushing toward Chile at the same time that inland travelers were following rivers into North America. As if drawn by the emptiness, these people went as far as they could.

Sites appear more quickly than they should if the people had moved and populated the way that modern hunter-gatherers do. Instead, they filled spaces like fast water, either by coast or inland. Todd Surovell, an archaeologist specializing in the Paleo-Indian period at the University of Wyoming's Frison Institute, wrote in 2000, "It is quite possible that the Americas were populated very rapidly by highly mobile hunter-gatherers." Whether they still carried an atavistic wanderlust from breaking out of Africa long ago, or they reproduced in unusually high numbers—an adaptation seen in pioneer species, such as coyotes or pigeons, which reproduce at an increased rate while populating empty regions—the first people moved at a swift clip. A 2016 study of 7,200 radiocarbon dates from early hunter-gatherer sites in Colorado and Wyoming backs up Surovell, concluding that these people "achieved long-term rates of population growth comparable to the rates of population growth achieved by prehistoric New and Old World agriculturalists."

Dillehay describes a slow coastal advance, but other archaeologists have seen it as a race, people who couldn't stop paddling, or reproducing, until they'd spread all the way down the coast. Surovell constructed a model to estimate the possible speed of Ice Age coastal migration, starting near Seattle (the maximum southern limit of the Cordilleran Ice Sheet) and reaching to the bottom of South America. He imagined a minimum population of 175 people within a thousand-kilometer strip of coastline and sent them leapfrogging south at a 3 percent growth rate. Each generation in his model could only move as far as two hundred kilometers before optimal density was reached, and new migrants moved yet farther south in search of new food sources. Under these controls, the fastest Surovell could get his people from Seattle to Monte Verde was 2,267 years.

Traveling from Alaska to Monte Verde does not have to take that long. Any boat travelers seriously questing for the edge of the earth could cover the span in a year or two and be back before their departure was forgotten. I spoke with a young adventurer, a kid out of Juneau, who traveled mostly by kayak, sometimes by bicycle, from his hometown in Alaska to southern Chile. By human power alone, Kanaan Bausler had managed to cover the exact distance traveled by the first people. His was not a sponsored expedition; no magazine articles or photographers went along to cover the adventure. It was simply what he wanted to do after college and a couple of jobs, much like me on the Yukon at the same age. This may be how the continents were first explored, small numbers out to see what they could see.

Kanaan Bausler's intended destination was Tierra del Fuego, the southernmost tip of South America, another thousand miles from where he ultimately stopped. Bausler, twenty-three at the time, was able to cover eighty-four hundred latitudinal miles along the Western edge of the Americas in two years. He passed from the glacial North to the glacial South, with bands of desert and rainforest between. Seeing him as an Ice Age proxy, I asked what the world looked like at this pace.

"In a sense, it was unremarkable," Kanaan said. "The change happened so slowly that each night didn't change that much. When you're constantly seeing something different, it's not an epiphany."

Kanaan's epiphany arrived not overnight, but in the last few thousand miles. Twenty months after leaving Alaska, passing beyond "the unbearable heat of Central America," Kanaan saw fresh snow again from his kayak in Peru. Down the coast of Chile, the moon kept turning until it was fully upside-down, and all his Northern stars were gone, replaced with new constellations. Here, the coast began to look familiar. Temperate forest came down to the shores, mountains rising, glaciers tumbling into rivers and bays among islands and fjords. It looked as if he had—somehow—returned home.

The place Kanaan decided to stop is, for me, an epiphany. He pulled over near the city of Puerto Montt. Not far from the city, a river runs to the sea, and it passes by the Monte Verde site. He had no idea of the archaeological significance of the region, but he ended his journey here. He had reached serrated islands, inlets, and sheltered waters. As much as he wanted to complete his original goal and reach the tip of South America, the feeling of homecoming was irresistible. He had to stay.

It is my belief—not through any archaeological evidence but because of the shape of the land and what Kanaan did when he got there—that the Monte Verde archaeological site was established because it reminded early boat travelers of their home. Whether their journey was a quick shot like Kanaan's, or something that took many generations, a complex expedition with grandparents and children in tow, people ended up at Monte Verde, what one scientist told me was the "Gilligan's Island of the Americas."

Kanaan left Juneau for South America at twenty-three. He didn't return to Alaska for another three years.

But he did return. And that matters: He was able to tell his family and friends about his journey. Unless the traveler returns, no knowledge can be passed on; no map will be laid out, no later migration of families will occur. Kanaan came back and told the people of Juneau what he saw. He told them he'd passed beyond countless horizons, and at the far end of the world he found a familiar place with unfamiliar stars and an upside-down moon. Who could resist such a call?

Did the first people imagine the shape of the earth? They might have known the general shape of the planet; they had enough information to compare the roundness of the sun and the moon to what was beneath their feet. Or perhaps they saw it as a platter of earth, a flat or perhaps slightly curved land around which sun, moon, and stars perpetually revolved, passing into the underworld before rising again.

Or maybe it was far more complicated, and with their extra 5 percent of brain volume, they took in the geography differently than we do. Those traveling down the Americas might have found a doppelgänger of the Northwest Coast in Chile and thought the world was folding over itself. They were returning to a similar place, but it was not exactly the same. I have traveled into the coastal ranges of Chilean Patagonia, and they feel like Alaska, their islands and glacier-horned peaks dead ringers for places I'd seen in Northern latitudes. Rowing a raft down the Rio Baker in Patagonian Aysén, coming out on its brackish gray-water delta, I could have been coming out of the Stikine River into the Inside Passage of coastal Alaska.

What comes to mind is the topological concept of a "Klein bottle," a single, nonintersecting surface that wraps around itself; if you traveled across it, you'd find yourself returning to the same point, only upside-down. Beaching their boats on the ragged coast of southern Chile, they may have been puzzled at their new but not so different environment, as if the world were a woven basket and you were moving through its circular warp and weft.

Kanaan Bausler made the trip, and I believe that sometime within the colonization history of the Americas an Ice Age individual must have done the same. Surovell's model for a 2,267-year journey from the coast of North to South America leaves out a crucial factor, something another archaeologist pointed out to me. I was visiting an excavation on a tiny island off the coast of southern Maine, where I rowed a white wooden rowboat to a dig site. The excavation at Smuttynose Island was bringing up the foundation of a 1770s Revolutionary War tavern, and below it were Paleo-Indian stone tools and weapons. The University of Southern Maine crew was led by archaeologist Nate Hamilton. Sitting along the trenches where his crew had found buttons, coins, a ceramic pipe, and, down deeper, diagnostic stone weaponry, Hamilton said he thought archaeologists are missing the point. He asked why people reaching the West Coast would have crossed the continent for Maine when there would have been plenty for them to eat and drink along the way, no practical reason to keep going. Why leave California? Why leave the fertile Great Plains?

"What's over that mountain, and that one?" he asked. "Adventure. It's one thing archaeologists don't credit enough."

Adventure is not in Surovell's model. How much D4 were Ice Age people packing? I go back to Diane Hanson's warning that scientists tend to make early people look like themselves. Perhaps people had more verve than we do today; they may have picked up the pace, felt compelled to move, pouring into empty spaces. Kanaan Bausler puts the journey from Alaska to Monte Verde at two years, assuming that you dawdled like he did, fishing and buying tacos along the way. If you were moving kids and grandparents down a coast, I'd add another year or two for safe measure.

So how long would it have taken for boat travelers from the Pacific Northwest to reach their mirror image in southern Chile on this Ice Age coast? A ballpark figure: two to 2,267 years.

Night by night the skies darkened down the outer rim of British Columbia. Stars came out, the first we'd seen in weeks. Coastal towns began to appear, more lights on shore, more boats in the water, more floatplanes in the air. The temperate world was opening, allowing for more and more life. Cruise ships appeared like ivory castles on the water, and humpback whales breached and sent up jets of mist.

On our final night we plied the backside of Vancouver Island. Lulled by the basso engine deep inside the ferry, the boys were finally asleep. At the far southern end of the island, which passed darkly on the starboard side, the fragment of an arm bone from an unusually large *Arctodus*, the short-faced bear, was found. Its chemical condition suggests that the bone, along with the rest of the carcass, was deposited in "a steppe-like grassland environment," now a forested island coast. The *Arctodus* bone has a radiocarbon date of about twenty-two thousand years old. Nearby, a mammoth leg bone was found, seventeen thousand years old. Further south on Orcas Island in Washington State, the bones of a twelve-thousand-year-old bison, *Bison antiquus*, were found, butchered by human tools. South of that, a 13,800-year-old mastodon skeleton was found on the shores of the Olympic Peninsula. The animal lay on its side, with a projectile made of sharpened mastodon bone embedded deep into its spine.

I quietly unzipped the rainfly and slipped out onto the ship's deck. Skies were partly clear, a handful of stars in the almost-black. Other travelers had bedded down under a nearby solarium. A few headlamps were still on. I saw a bearded, rough-looking young man writing in a journal. I'd noticed him when we came on board in Wrangell; he had the shell-shocked look of someone who had wandered out of the wilderness and wanted nothing more than to return to it. This was our last night, and I wanted to say hello. I grabbed a bottle of red wine I'd bought in Ketchikan, and headed over.

He said his name was Josh. He thanked me as he held out a metal cup and I poured it half full, before doing the same with mine. He said he'd just finished a solo thousand-mile kayak voyage starting in Bellingham, Washington, ending in Glacier Bay, Alaska. Now he was on his way back home, his kayak stowed below with all the other vehicles. If he went again, he said he would take longer.

I told him I'd always thought of going north to south, and not the other way around. His journey seemed almost against the grain.

"It's not like it's downhill." Josh laughed.

I poured more wine, asking if it felt like a super-human feat to go that far alone. "It could be done by almost anyone with half a sense of adventure," he said. "Really, I think people don't know what they're capable of until they get out."

Josh told me how free he'd felt, how good it is to go on a journey and explore the world around you with all your senses. And how strange it is to come back to engines and exhaust and busy sounds of people. He told me that a month wasn't enough.

"I'd recommend three or four months," he said. "However long it takes, the longer the better."

We drank half the bottle. I said good night and left him the other half to curl up with, a welcome back to the human race.

In the morning, I woke to the blast of the ship's horn. I'd accidentally slept in. My head ached. The boys were out cold. A voice came across the PA saying that in half an hour the *Columbia* would be docking in Bellingham.

"Boys, boys, it's time to go, we've got to hustle."

They moved slowly and made ungrateful noises.

I shook them and said, "You have thirty seconds, then I'm pulling you out."

The older boy came up snarling, and the younger, with stitches in his forehead, burrowed deeper into his bag.

It was a sunny day in Bellingham. The small coastal city came around us and the ferry ran its engines in reverse. The ship shuddered against wooden pilings.

Our tent was still up as campers cleared out of the solarium, and a purser came around making sure passengers were leaving. Shoving and jamming books, playing cards, and sleeping bags into packs and pouches, I snapped my fingers, telling the boys to pick up their pace. Breakfast was whatever I unwrapped and got into their mouths. Blood sugar dropped faster than I could counter.

The interior spiral staircase of the ship wasn't large enough for all the bags and packs at once. The boy in front bickered with the one in back, so much gear draped awkwardly around us that we were more like luggage than a family. Someone cursed loudly. It may have been me.

I stomped onto the metal gangplank that connected the *Columbia* to the continent. I had been planning an auspicious entry. Maybe I'd release a pinch of ochre to the breeze while saying out loud, *Welcome to the land of beavers that stood six feet tall and Columbian mammoths from here to Mexico. Welcome to Devils Tower, the Grand Canyon, the Ozarks. Welcome to the Colorado River, the Gulf Coast, the Rocky Mountains.*

I said none of this because straps were digging into me. At the far end of the gangplank stood Regan. The boys ran to her while I came behind them, carrying almost everything. Red-faced, I reached concrete and dumped gear on the ground.

This is how we arrived in the New World, *dammit.*

5

PLAYGROUND OF GIANTS

45,000 TO 15,000 YEARS AGO

In 1781, Thomas Jefferson learned of a man who had returned from what was then the frontier, describing bones that came from creatures bigger than any known living land animal. Jefferson wrote that "a Mr. Stanley, taken prisoner by the Indians near the mouth of the Tanissee, relates, that, after being transferred through several tribes, from one to another, he was at length carried over the mountains west of the Missouri to a river which runs westwardly; that these bones abounded there, and that the natives described to him the animal to which they belonged as still existing in the far northern parts of their country; from which description he judged it to be an elephant."

In Jefferson's era, mammoths, or at least the bones of mammoths, were seen as the remnants of creatures from before the Great Flood. He'd heard of a find at Big Bone Lick, Kentucky, where many mammoths had left their skeletons in place, as if they'd died during drought or hard times, which they had, though not in some impossibly ancient time. The bones were still bone, not yet mineralized into actual fossil. Jefferson did not know that human artifacts of equal antiquity would be located among the mammoth remains of Big Bone Lick. He had his suspicions, though. And Mr. Stanley's report intrigued him. "Running westwardly" from the Missouri would have put the man on the west side of the Continental Divide. Jefferson thought that if creatures like mammoths still existed anywhere on

Earth, they might be somewhere out West, lost in the unexplored country of his people's new continent.

In 1803, Jefferson sent Meriwether Lewis to Big Bone Lick to acquire some of these mysterious remains, hoping they might help him better understand the place of man and beasts on God's planet. The following year, pleased with his success at collecting mammoth remains from the state of Kentucky, Jefferson sent Lewis and William Clark to the head of the Missouri for an expedition into the American interior where the president hoped he might hear something more about mammoths. The expedition returned with a tale of a vast and well-inhabited continent, many languages and incoming rivers, but no sign of living proboscideans.

More than a century later, James Teit, an anthropologist who gathered ethnographic stories in the Northwest alongside the famed Franz Boas, who later became known as "the Father of Anthropology," published tales from the Kaska people of the northern British Columbian interior. In a footnote to his 1918 study, Teit wrote that he heard of a "very large kind of animal which roamed the country a long time ago . . . It was of huge size, in build like an elephant, had tusks and was hairy. These animals were seen not so very long ago, it is said, generally singly; but none have been seen now for several generations. Indians come across their bones occasionally."

A Penobscot story from Maine recorded in 1934 in *American Anthropology* tells of a hero figure called Snowy Owl who "was searching for a wife far to the south. He noticed that the watercourses were drying up and followed up a valley to seek the cause. He saw what seemed to be hills without vegetation moving slowly about. Upon closer scrutiny he saw that these masses were really the backs of great animals with long teeth, animals so huge that when they lay down they could not get up. They drank for half a day at a time. Snowy Owl went on and after many adventures secured his wife. Then he returned to the place where the animals had their 'yards.' He cut certain trees upon which the monsters were accustomed to lean at night so that when they did so the trees would break. Thus the animals fell upon the sharp stumps and Snowy Owl shot them all."

Could this be a story about hunting megafauna into extinction? Oral histories are not hard data. Scientists turn to them for much-needed context, for a voice to give texture to artifacts and papers, but all stories change over time, bending and bleeding into new traditions. Indigenous mammoth tales could as easily be reflections of scientific discoveries, people adding back into their story what they have learned of their own past. When researchers and collectors came from far away to dig up tusks and giant skulls in Siberian permafrost, Ugric speakers living there might have given these long-gone beasts a name: *earth-horn, muw-xar.*

Delving deeper than oral tradition, returning to original bone, I followed Dr. Sue Ware into a storage hall in the Denver Museum of Nature and Science where boxes were stacked, and limb bones and vertebrae rested on shelves. A pair of reassembled dire wolf skeletons was posed running across the top of a storage cabinet, as if taking over the place.

This part of the museum was originally the Ice Age Hall. I knew it from when I was a kid living in Denver, Quaternary murals scrolled across the tops of the walls, skeletons assembled upright, mammoth and mastodon returned to their original heights. The hall was now closed to the public and filled in, stuffed to the gills, temporary desks and offices made between crates where space allowed.

Ware walked through exhibits turned into storage, pausing frequently to point out some bit of knowledge or specimen, a beautiful short-faced bear skull, the pelvis of a giant ground sloth that, as she noted, could comfortably "seat two." Ware and I are old friends. We've messed around in the museum together from way back, opening predator jaws and peering inside for sheer amazement.

Ware studies large fossil and modern carnivores, calling Ice Age North America "the playground of giants." She has a mischievous grin when she talks about the bloodbath of the Ice Age. Her dissertation is on pack behavior among extinct American dire wolves.

When I was a kid and came to the museum on school trips, I'd drop coins into the mouth of a hard plastic sabertooth cat just inside the entrance. I remember the cat's head and shoulders on a wooden box like a bust, mouth wide open, saber-teeth curved. Coins disappeared down the dark hole of its throat and from inside the box a roar would come out.

I could have fed that cat coins until all my lunch money was gone.

The voices of classmates echoed off limestone floors and columns holding up the ceiling as we filed in. Once inside, teachers could scarcely keep us together. I was on a mission. I passed the tank-headed Eocene herbivores and other pre-human oddities. I liked the dinosaurs enough to have a geeky glossary of their Latin names in my head, but the prehistoric animal I believed in and was enthralled by, the gigantic beast that human beings actually *knew*, was the mammoth. I stood beneath its curved golden tusks, the skeleton upright above me, *Mammuthus columbi*, the Columbian mammoth, the thing Jefferson hoped to find alive in the wild American interior. Its skeleton was bolted to the ceiling with metal rods, as if all it required was the word to break free.

My return to the museum this time was for the same species, *M. columbi*, yet another entity named for Christopher Columbus and his unapparent discovery of the New World. I was inquiring into the remains of the front quarter of a Columbian mammoth found in a clay layer at the bottom of a pond near the ski town of Snowmass, Colorado. The mammoth's remains are forty-five thousand years old, its age determined by dating organic matter buried in the peaty mud around it, and eleven rocks were found intermingled with the bones. The rocks are puzzling because they are found nowhere else at this site except on top of the front quarter of the mammoth. It's hard not to wonder if they were intentionally placed there.

The site was designated as paleontological only, not archaeological. Museum scientists instructed researchers to speak of it in these terms, saying there was no definite human component, no scrapers, spears, or knapped stone. But the rocks are peculiar. Big enough to pick up and carry in human hands, they are each roughly the same weight and size, as if they were chosen.

A similar phenomenon has been recorded at glacial pond sites in North America and Siberia, rocks used to hold proboscidean carcasses underwater, thereby both preserving the meat and protecting it from scavengers. In Wyoming's Agate Basin, a pile of articulated bison parts with other bison bones is believed to have been a frozen cache. Dr. Daniel Fisher, director of the Museum of Paleontology at University of Michigan, studied megafauna-caching techniques in the Great Lakes region; he has tried it out himself on a deceased draft horse, which he butchered with stone tools and anchored under ice. He reported at a Society for American Archaeology meeting in Minneapolis in 2014, "As long as ice remained on the pond, the meat stayed essentially fresh. By June, the meat had developed a strong smell and sour taste, but still retained considerable nutritive value."

In 2015, Fisher excavated a nearly complete mammoth in Michigan. Its skull, with two intact tusks curving elegantly away from the head, was hoisted out of the ground on straps from the bucket of a backhoe. Stone flakes and the arrangement of the bones suggests that humans were involved, butchering the animal in one place and separating the meat elsewhere—caching it in water, he thinks, for later use.

Fisher was called in to excavate the Snowmass mammoth in case it turned out to be the same phenomenon. He came away thinking it was, but the evidence remains too thin to say for sure. Many of his colleagues are not convinced; eleven rocks are not enough to prove human activity so long ago.

The Snowmass mammoth is known as the "clay mammoth" after the lens of clay in which it was found. When the discovery of the animal and others just like it was first announced, I drove over with my family to see mammoth and mastodon long bones, tusks, and teeth laid out on tables fresh from the museum's excavation. Children were allowed to touch some of the dark remains, bones saturated from resting in the muck beneath a mountain pond. It was a combination of time travel and ancestral faunal worship, lines of people from surrounding communities filing in to see the discoveries. More than thirty-six thousand bones came out of the excavation in a feverish sixty-nine days as museum excavators worked ahead of construction crews on the floor of a drained reservoir being deepened for water storage. In all, parts of four individual mammoths and thirty-five mastodons were excavated from this fifteen-acre mud pit. Each had died at a different time, some tens of thousands of years apart, making this not a kill site but a death site, the result not of hunting but gradual mortality in a place where proboscideans spent a lot of time.

A bison skull had been found in dark loam along with at least ten others of the same species. Ware showed the skull to me as we passed through an aisle, and it was as big as my chest, long horn cores protruding to either side. This was *Bison latifrons*, known to have horns stretching more than six feet from tip to tip.

"That's what we call a holy-mother-of-God bison," Ware said.

Latifrons grew so large—possibly the largest bovine the world has ever seen—partly because of the wealth of Ice Age forage. It was also

Skull of *Bison latifrons*

a size that would have given pause to *Panthera atrox, Smilodon,* and dire wolves. Modern bison are thought to be smaller as an evolutionary reaction to warmer climates and the need to run longer distances, possibly to escape human hunters. The Ice Age was a different era for megafauna. It was stand and fight, defend yourself and your turf, the bigger the better.

Most archaeologists are now willing to concede, conservatively, that humans were south of the ice by fifteen thousand years ago. But a quartered forty-five-thousand-year-old mammoth stored in cold water by humans at Snowmass would throw those dates off the chart. The clay mammoth falls into the category of Richard Morlan's forty-thousand-year-old shattered mammoth bones found in Old Crow Basin, Yukon Territory. Archaeologist Steve Holen, former curator of archaeology at the museum in Denver, now running the Center for American Paleolithic Research, has identified ten incidences of possible human-shattered megafauna bones from the La Sena site in Nebraska and the Lovewell site in Kansas from around the last glacial maximum. What is now the semi-arid steppe of the dry High Plains was then sagebrush and grasslands mixed with the occasional spruce or pine. Holen's finds in this former prairie suggest earlier waves of people than most scientists will allow. In one case, he reported, "Impact points on mammoth and large ungulate limb bones show

how the bones were broken for marrow extraction and the production of bone tools and choppers." People were not only gathering marrow, they were making tools out of bone and carrying them away, leaving no other sign of their passage. It was the march of bone-smashers from the north, assuming they were people who actually existed.

In 2017, Holen and his colleagues analyzed bone smashing from one-hundred-and-thirty-thousand-year-old mastodon remains found at the Cerutti Mastodon site in San Diego. These came with abraded stone cobbles, and the bones are arranged in a somewhat orderly fashion. The site is hard to mistake for anything but human, though many scientists don't agree, finding it too old to fit current models. It may, however, fit a more realistic model where humans and even earlier hominids arrived not in one single time period, but many. The river cobbles found at this mastodon site were not near any river that could have rounded them, nor were other cobbles found in the area. They are out of place, and their pitting and abrasions suggest that they were struck against something solid in a purposeful manner, with a certain amount of focus. A uranium-thorium dating technique put the mastodon bones at 130,000 years old, which, if it holds, would have humans here in the Upper Paleolithic.

People, or early hominids, may have made an appearance in North America thousands of years, even 130,000 years, before foragers first carried stemmed points down the coast, or the makers of Dyuktai weapons and microblades gathered across Beringia, seeking a way through Alaskan ice. Whoever these early bone-smashers might have been, they arrived and then vanished. It seems the harder and deeper science peers, the older the firsts become. We want to know where we all came from, what oldest flickering thing in human history we can reach back to and touch. By every thousand years the record grows more frail. Bones are searched through, scanning electron microscopes used to study cut marks on mammoth remains. Holen's finds do not establish humans in North America at these far-end dates, but they are a knot on a string to follow into the dark.

At Snowmass, the tooth of a camel and foot bone of a horse also turned up in the dig, along with parts of three different giant ground sloths. One bone appears to have been chewed on by a short-faced bear, possibly scavenged. Ware showed these to me one at a time, the knuckles in a box, a piece of vertebra on a shelf, a wing of a shoulder bone that would have covered most of a dining room table.

To say humans were at the Snowmass site forty-five thousand years ago requires considerably more evidence than eleven similarly sized rocks. Death rituals have been widely observed among elephants, which place dirt and branches carefully on their dead. Perhaps mammoths, being of greater heft, placed boulders on their deceased instead.

The going theory, however, is that these rocks fell from a mountain, rafted across the pond on ice, and happened to melt out directly above a sunken front quarter of a mammoth, falling neatly into place, no human agency required. Ware gave me a look of incredulity. "Really?" she said. "Is it so difficult to imagine people were here?"

The first people would have been walking into a brawl, whether they got here as far back as 130,000 years ago or just 15,000. Studies of damage to carnivore teeth and jaws from the La Brea Tar Pits near downtown Los Angeles have shown high-impact living among Pleistocene megafauna, the sabers of sabertooth cats snapped in two, skulls of predators fractured and healed from blunt trauma. From the oily, sticky pits, UCLA paleontologists Blaire Van Valkenburgh and Fritz Hertel examined skeletal remains of dire wolves, American lions, and sabertooths, and concluded that these carnivores "utilized carcasses more fully and likely competed more intensely for food than present-day large carnivores."

Ware doesn't fully agree with this assessment, saying there is no way of knowing whether dire wolves or other predators utilized carcasses more fully, or whether they competed more intensely for food than present-day carnivores. "All we can say is they broke teeth, had injuries consistent with hunting megafauna, and died by entrapment," Ware said. "Modern wolves that I have studied are competitive, aggressive, and have a definite carcass protocol that utilizes a carcass in an efficient manner. We just don't know about fossils, we can only talk about their bones."

Pleistocene coyotes were maybe twice the weight of their modern relatives, their skulls larger, teeth more precisely designed for shearing flesh, molars showing wear that suggests extensive large-bone eating. They had the same damage patterns and frequency as modern gray wolves, pointing to a larger target prey size than modern coyotes. Everything was kicked up a notch, similar patterns appearing in the fossil beds of San Josecito Cave in northern Mexico just south of Texas, and the coal-black tar traps of Talara, Peru. Humans were walking into one of the biggest meat-and-bone shows on Earth.

The last peak of the bone damage appears to have occurred around fifteen thousand years ago. Wolf species at the time had three times more mouth damage than animals just three thousand years later, around twelve thousand years ago, meaning competition might have been fiercer, predators in greater abundance, and more and larger carcasses were scavenged and eaten. Ware speculated that humans were most likely viewed as a food source by Pleistocene predators.

Ware researches extinct dire wolves, an American animal similar to the modern wolf but larger in every way, notably in the skull. Although she is trained as a physical anthropologist, she is not an archaeologist. She prefers studying the relationships between humans and the animal kingdom. Her dissertation showed that dire wolves once lived and likely hunted in packs, much like modern wolves. She determined this by studying broken and healed dire wolf bones from the La Brea Tar Pits. Some with terrible injuries lived long enough for their bones to begin to heal or fuse back together. The animals did not ultimately die of their injuries, but rather of entrapment in the tar pits. Her conclusion was that they needed some help to survive their injuries, the support of a pack to nurse them back to health.

Ware imagined sophisticated communication networks implemented as dire wolves closed in on a mammoth, a horse, or a human. Her "boxy" canids would have brought down their prey by hamstringing them and ripping at their flanks. A sabertooth attacks differently, slashing arteries and organs so the animal will rapidly weaken and bleed out. The hamstringing technique would have been more carefully orchestrated with more animals involved, pack hunters.

I'd gone into the pathology collection Ware used for her dissertation at the Page Museum in LA. The dire wolf bones she studied had soaked in tar so long that they had a blue-black sheen, as if made of coal. The pathology collection was a gallery of dark grotesqueries: burls and healed breaks, femurs and skulls that look like Swiss cheese or beehives. A disturbing thing to witness, it is a hall of pain on shelves that the public never sees. I looked at knots and breaks showing where canids had almost met their end, but were kept alive by their packs.

During her dissertation, Ware and I spoke frequently. She confided that she was spending a lot of time thinking about what happened here, how the wolves must have dominated the landscape, and how they died. Going through tortured bones, how could she not wonder what La Brea must have sounded like as the animals became trapped, or what the smell was like in the tar as corpses sank into the mire?

At the time, Ware had just been diagnosed with a rare bone disease, a form of osteomyelitis that could cause damage uncannily similar to the kind that afflicted the dire wolf bones she was studying. An infection can enter the bone through the bloodstream or migrate from nearby tissue. Her disease could not have been caused by the Ice Age bones she studied, simmering in oil and tar for tens of thousands of years. But it was the same kind of infected osteological wound, her bone appearing hazy on the X-ray, as if it were vanishing, too full of holes to bounce back a sharp image.

Her foot was ultimately amputated before the disease spread, taking off most of her leg below the knee. Ware, who had hoped to include her foot—after proper preparation—in her osteology collection, asked if she could have it back to send off for processing. But it was not to be. The specimen had been incinerated.

Now she walks with a prosthetic. On her replacement leg she illustrated the face of a wolf.

Walking slowly, still learning to balance herself, she paused in front of a long bone from the clay mammoth. Darkened by layers of peat and bog water, the bone was the color of whiskey and shoe polish.

"Those scratch marks." she pointed. "Do they stand out to you?"

I could see what she was talking about on the rim of the bone, a couple of strokes possibly from a stone tool, possibly not. Like most finds from this site, this one had been naturally roughed up, scratched and gouged.

Researchers examining these marks on a microscopic level have concluded they weren't caused by people, but by grit shifting in the matrix, maybe trampling by other animals. But some of them catch your eye. I asked about one rib that had clearer markings. I'd seen pictures of it, scratches looking like deliberate, repetitive motions, as if a single blade had been pushed in the same direction forty-five thousand years ago in the Colorado Rockies. Though the microscope said it was nothing, my eye couldn't help seeing rapid sawing motions through muscle and fascia. She said the specimen was out for another study. Maybe it would come back with a different conclusion this time.

First people were interesting but they were not Ware's primary interest. She was more intoxicated by what those people may have encountered. She saw in this mammoth bone an alpine valley in the middle of nowhere, the continental interior of the New World. The Roaring Fork River through Aspen, Colorado, came out of glacier-clad mountains. Up one of its tributary valleys was a pond fringed with fir and spruce, hidden at the far end like a jewel. Sunlight landed in the back of this U-shaped valley where mammoths, mastodons, and giant *latifrons* bison gathered in marsh grass. For a moment, the giant cats would have backed away, unsure of these spear-bearers and their dogs. Mastodons would have lifted their heads, wondering at the commotion.

6

EMERGENCE

16,000 TO 14,000 YEARS AGO

In the oaken hills of central Texas between Austin and Waco, a spring rises along a tangled fence line, watercress waving downstream. Water comes up clear and clean, like diamonds running between willows and meaty cottonwood roots.

Even on dry years, Abbott Spring keeps a good flow, two to three liters per second from deep below the head of Buttermilk Creek. A short walk along the creek reveals dove-gray pieces of chert everywhere. You can't walk without stepping on them. The kind of rock is called Edwards Plateau chert, a local outcrop popular for toolstone since the late Pleistocene. I know a modern knapper who comes here to collect rock to take home and fashion into stone projectiles. It's one of his favorite toolstones to work with.

An archaeologist, the only one working the weekend shift at what is known as the Gault Site along Buttermilk Creek, showed me where butchered parts of a mammoth had been excavated and filled back in. Nearby was the site of a horse, a Pleistocene bison. Some of North America's oldest, most well-stratified human dates have come from this site, dates as far back as 15,500 years ago. This was one of the earliest documented gathering places on the continent, material generated by people coming back over and over.

Buttermilk Creek is a hunting, butchering, and stone-gathering location holding the remains of so many stone tools left across so many thousands of years that the ground seems to be as much tool-flaking as actual dirt. Charcoal, bones, and knapped rock from the 15,500-year-old horizon up to the historic Comanche people constitutes earth.

The Gault School for Archaeological Research out of Austin, Texas, has a central dig under a white tent that looks like a religious revival in a grassy field between oak bosques. Fifteen thousand years ago, this land was twelve hundred miles south of the ice cap. It was temperate, environmentally similar to what it is now. The spring would have flowed about the same, toolstone being struck, crystalline sounds ringing through oaks, the music of the first people. A camp near a spring at the head of a creek with good toolstone would have meant the scent of smoke drifting through woods, and you would have heard human voices, languages brought down from Siberia, or perhaps carried over from the Iberian Peninsula, beyond the Atlantic, as some researchers believe.

The oldest clear sign, so far, of architecture in North America was discovered just across the field from the tent. The floor plan indicates a rectangular layout, the distribution of artifacts made of stone and bone revealing an outline of walls, an entrance, a cooking and eating area. A floor plan as defined by archaeologists contains the shadow of living. If I'd ducked into this rectangular hut all those thousands of years ago, I'd most likely have been overwhelmed by the smell. To them, it would have been the scent of home: wood smoke, meat and skins, and family—a refuge from woodlands full of mammoths and sabertooth cats.

Home is a shape, a space, and for the people at the head of Buttermilk Creek it had square walls. The structure was dug into the ground with a fire pit inside, its floor plan divided into tasks, an orderly arrangement.

Similarly laid-out Paleolithic footprints have been found in Europe, rectangles dug in with fire pits and artifacts in the same positions, entrances designed the same. This was pointed out to me by Michael Collier, one of the senior archaeologists at the Gault Center in Austin. Collier opened plan-maps from these European sites, showing shapes and artifact distributions similar to what has been excavated at Buttermilk Creek. He said what puzzled him was that the Texas site he excavated is older than those in Europe.

I asked, "You mean they could have been moving from here to there?"

Collier leaned back from the maps and said, "You said it, I didn't."

The question of where people went, when, and how could find answers in a pit like this; tools tell origin stories, bones of animals speak to meat and fire.

Near the entrance of the tent at Buttermilk Creek is a signpost with three arrows pointing different directions. One is marked *Iberian Peninsula*, the other *Siberia*, and the third *Monte Verde*, in Chilean Patagonia. These are three key locations for the nascent Americas, the Iberian Peninsula being a reference to a possible arrival of people from Paleolithic Spain, Siberia pointing back to the land bridge, and Monte Verde referencing an anomaly on Chile's southern coast. The signpost felt like a collective sigh from the archaeological community: *We're not sure where everyone came from — maybe everywhere.*

A Columbian mammoth would have been dappled in shadow, standing still, waiting out the afternoon head-deep in the shade of the oaks. Its tusks would have been raised high off the ground. The slightly smaller mastodon has tusks that curve gently forward. The Columbian mammoth's tusks bow outward and then curve back in, almost touching, like a Valentine heart.

The big white tent through draping plastic doors was cavernous, a lab and temporary artifact storage focused around an excavation pit almost twenty feet deep. Fans whirred and a generator ran outside, pumping water from the excavation so it wouldn't fill up like a swimming pool from groundflow.

A wooden stairway had been built from fresh lumber, the wood unfinished, everything screwed together so that this place could eventually be taken apart again, the hole backfilled, grass grown back into a pasture where livestock would graze, knowing nothing.

The uppermost excavated horizon was the size of a living room. Down through tiers, the pit became smaller, passing into older dates, arrowheads and small points belonging to post-Columbian Indians, Comanches. Below them were mound builders, woodland hunter-gatherers from a thousand years ago, people who would not have known of Europe. Another few steps, and I was at eye level with layers of Archaic big-game hunters from the cooler interior of the Holocene, six thousand years ago. Down the wooden stairs, flakes and bits of charcoal stuck out of walls, the leavings of people returning over and over for the goodness of this spring and its Edwards Plateau chert. Below them were Paleoarchaic hunter-gatherers who also knew of this chert, learning of it from those before them, the Pleistocene people, the mammoth eaters.

The stairs continued down past nails tied with bright string marking study grids, damp dirt walls becoming soaked the lower I went. The water table came up around me, bones of horses and mammoths found in layers, part soil and part water. At thirteen thousand years ago, I was in the Clovis age, the rise of the great mammoth hunters, long thought to be the first people in North America. Their unique stone-tool technology swept the continent like wildfire. I descended deeper into human occupation, fourteen and fifteen thousand years old—the pre-Clovis layers of Buttermilk Creek, which have produced 15,528 artifacts in the form of broken pieces of tools, rock flakes, projectiles, and worked bone, proof that Clovis was not first.

At the bottom of the pit, a wood plank was up on blocks above standing water. The plank bowed under my weight, touching the oldest, muddy human horizon, last of the worked stone, sterile soil below. The bottom date at Gault has reached 15,500 years old, a solid time frame for human arrival in this part of Texas.

Clear plastic bags were in a bucket, each numbered, each holding a rock flake that had been found in this deepest horizon. I thumbed through to the bottommost artifact, a humble chert core knocked around the edges to get it down to a workable carrying piece. I opened the bag and slid the worked stone into my hand. It was a discard, something the toolmaker had thrown aside. Maybe the particular rock wasn't liked, had too much patina on one side, or was too thick in the middle. It could have been picked up and worked again, but it was forgotten. Almost as big as my palm, it was heavy, and sharp enough it could have cut flesh. My hand wrapped around its cool surface, slick and gritty. It felt amniotic, still wet from birth.

I spoke with a Navajo woman, Nikki Cooley, who graduated from Northern Arizona University and works in the Grand Canyon as a river guide. When we talked about where we each came from, she told me, "I firmly believe we did not migrate here. Maybe we crossed the Bering Strait when we were hunting, we probably went there and back, but I don't believe we came from there. We came from here: First Man and First Woman."

Instead of the scientific story, Cooley believes that the people came from the ground—her people's story. She told me, "My dad always says, 'People say, I come from Page, or I come from Flagstaff, when I don't come from any place but earth.' He picks up a piece of dirt or a rock and says, 'I come from here.' I have friends who are scientists and they talk about it. I politely listen, but I disagree."

Not everyone thinks this way. A young Navajo man I know out of Page, Arizona, doesn't care for fundamentalism. He tells me the earth is round and his people originally came from Asia. Disparities in belief are common in any human society.

The belief that the Navajo—a people who call themselves Diné, "the people"—and many other Native American groups arose through the ground, often chased by a world-ending flood, is ubiquitous. In Diné tradition in northern Arizona and the Four Corners area, the people came up from the ground like corn. When they first reached this world, known as the Fourth World, they found it covered in water and occupied by monsters. They had to unleash heroes to kill the monsters. In my mind, this means Ice Age lakes and floods when the land was ruled by megafauna, and the heroes came with spears made of rock or bone.

A Diné singer named Buck Navajo told me, like Cooley, that they did not come from Asia. We met in his northern Arizona hogan on a chilly November morning, last gold leaves dropping off his cottonwoods. The old man said that there was no land bridge in his people's history. His hogan was a domed mud hut with corbeled wood for a ceiling, similar in size and shape to the earliest known shelters in Siberia and Alaska. I asked when they arrived to take up sheep and corn and finally call themselves Diné.

The singer didn't speak English, at least not to me or to the companions who'd come with me. We were wanting permission to trek around the base of nearby Navajo Mountain, so we brought the old man ground coffee, grapes, apples, and a couple of twenty-pound sacks of Blue Bird flour. This is how it's done: Come with groceries, and be polite. He said to ask anything we wanted. I had one burning question: How did his people arrive in the Dinétah? I wanted to understand the ancestry of this place. I thought there'd be an oral history of migration, a cultural memory of coming down from the Arctic long ago. The old man's grandson stood over a half-barrel wood stove and interpreted as I went on spilling out my data: a cave full of Athabascan sandals found on the Great Salt Lake Desert near the top of Utah, signs of people coming from a colder climate into the desert. There were direct linguistic connections from the Arctic to Arizona, including a branch of Buck's own language still active in Central Siberia, where it appears to have shown up later—an American-born language washed back.

I was digging myself deeper into a hole, thinking like a colonist. If I was new to the place, wasn't somehow everybody? It was as if I had come here with special news for him, and all we had to do was break the language barrier and admit we were telling the same story, science and legend saying the same thing. The response was again, *No, they had not migrated to get here.*

I must have been misunderstood, I thought. We were lost in translation. Na-Dené is a famously complex language family; the Navajo tongue was used as a code in World War II to throw off Japanese military cryptographers. Vowel length changes the meaning of a word. If a vowel is expressed with a nasal tone, the word's meaning changes again. Even in English, the question *how did people come to be on the land* could come off as *how did they come to be a people?* That's where I thought we were hung up.

Buck nodded slightly and spoke to his grandson. Without a hint of rancor, as if the words were plain as day, the grandson told me, "There was no migration. We didn't come from somewhere else. We came from here, from the ground."

The old man gestured down with two fingers, which were aged and crooked from work. I looked at the ground, red dirt, Moenkopi hardpack. I finally understood him, at least as much as I could. The ground is where scientists go to find artifacts: Fire hearths, weapons, bones, and stone flakes come from deposits from so many millennia ago that the people must have emerged in a world of floods and monsters.

There is no clear geographic order to how any people first appeared. They seem to have followed rivers, but the sequence is hobbled because so much archaeology is lost. To scientists, it looks as if the lights came on in a house at night around fifteen thousand years ago. Texas, Oregon, Florida. Closet, attic, basement. If the appearance of bone-smashers forty thousand or so years ago was the fishing out of keys from a pocket to open the front door, this was throwing down the newspaper and putting up their feet: home.

To find the likely front door, come down the Northwest Coast below the ice and follow the big rivers in. These rivers were highways rich with salmon leading into the land through mountains and valleys opening into vast interior watersheds, first the Columbia River as you head south and then the Klamath. The mouth of the Klamath flows into the Pacific in Northern California through the Cascade Range. Nearer its headwaters, above Klamath Falls, it becomes marshes and lakes. These continue up to two forks of tributaries, the Sprague and the Sycan, and finally the headwaters and a crest on the east side of the Cascade Mountains looking over the edge into rolling country thousands of feet below, the start of the American interior.

Summer Lake Basin lies below in south-central Oregon, the lake now shriveled, appearing otherworldly, and anoxic with heavy mineral concentrations, not a place where people go fishing. The water body would have been significantly larger and ice-free on a spring day in the late Pleistocene, muskox grazing its edges, five-hundred-pound *Panthera atrox*—the giant American panther—perched on rock outcrops overlooking the herds.

Dark stumps and branchless pines from old wildfires stand at the crest now. A cool, damp wind whispered through the boughs of the ponderosas that haven't burned. Over the edge, several thousand feet down in the high desert, this vapid, kidney-shaped lake nests inside a shadow of its former self. From the crest I spotted a dark point of rock on the far side of the lake, far enough away that even through binoculars it still looked meager, like a charcoal smudge. It was once a peninsula reaching into an Ice Age lake. Now it's a lonely point of dry land holding a number of rough, shallow rockshelters, Paisley Caves. Dried human feces, dating to 14,300 years ago, have been found buried under layers of dust and debris in these caves. Along with human coprolites, archaeologists found long-tailed obsidian projectiles similar to those used on the coast of Siberia and Japan all the way down to Monte Verde in Chile, signs of coastal people arriving and moving inland via big rivers spilling from Oregon, Washington, and California into the sea.

I drove down from the crest and stopped in the town of Paisley, Oregon, population 240, where I asked directions. At the bespangled

Pioneer Saloon, open in the middle of the day, the bartender looked out the window at my rental car. She told me weather was coming and that the roads would be "slick as snot."

On the wall across from the bar hung a picture of Dennis Jenkins from the University of Oregon, who headed up work on the nearby Paisley Caves. It was taken in the field, a sunny day below the same point of rock I saw from the crest. At the bottom of the photo was written, "Dr. Poop." I asked the bartender if she knew him.

His crew came in every summer, she said. "He's our celebrity, the Indiana Jones of poop."

I finished my beer and got moving. Under growing storm clouds, winter not fully rendered into spring, I drove through sage-covered hills over bumps and ruts, places I had no business taking my little two-door. I didn't take the roads the bartender recommended, wanting to get up high first, have a look at this territory. The roads wormed up, over, and around draws and flanks of buttes where I winced at the percussions against the undercarriage. I parked on the treeless flanks of a little mountain range populated with dried-up stock tanks. Sage waved through basins and scarps of earth rising and falling in tidal, tectonic rhythm.

Spruce groves and grassy parks had occupied this area in the Ice Age, aspen trees twisting out of cracks in the boulders, green buds softening for spring. It wasn't that much colder then, between fourteen thousand and sixteen thousand years ago, but it was wetter, lusher, patches of temperate steppe frequented by horses and muskox. By the time Paisley Caves were in use, the height of the last glacial maximum long gone, six thousand years in the past, massive ice caps melted at a quickening pace, which meant fresh water everywhere; waterfalls, lakes, and rivers reaching their height, permafrost retreating from the ground, grasses proliferating. Megafauna had ample browse, meaning meat for predators.

Clouds streamed off the Cascades. Dabs of sleet pegged the ground, hitting my shoulders, nicking my cheeks. For time travel, stormy weather is best. The dry West can be nothing but sunny, everything cut into slices of shadow and light. On a day like this, sunlight turns into color and shape as shadows richen. Ghosts stir from the ground, silver-tipped rain on sage.

A piece of knapped obsidian lay at my feet, exposed by weather. I rubbed pale soil from it, spitting on the rock, thumbing off the mud to clean it. It was a scraper, a small cutting tool with quick edge work. I held it up and gray sky seeped through its glassy fringe. Smoke-heated volcanic rock, common around here, it had been worked into a handy, portable tool. A flint knapper I know said to me, "Obsidian is a whore." A learned nod to its easily sharpened edge that dulls quickly, in no way a long-term relationship. Flipping it over and back, I could see hammer blows, first with a harder rock to knock off most of the junk, then the butt of an antler for finer flakes to sharpen the edges. It wasn't a museum piece, and who knew what age, a generic tool. You could find ten of these in a day out here. On intermountain shorelines from here to the terraces above the Great Salt Lake in Utah, you'll see lithic scatter that landed like rain, the ground winnowed down to human enterprise: tool flakes, bits of broken projectiles, and abandoned cores of glassy rock.

Stone Age artifacts are much easier to find in the sage desert than in a forested place like New England, where soil is scarcely visible beneath fallen branches and leaf litter. Many finds in the Northeast come from toppled trees with their rootballs exposed, rooms of soil walloped up from the ground, bringing up with them spearpoints, scrapers, and toolstones. The West is different, drier, more exposed. It is what is known as a "deflating" landscape where sediments tend to blow or wash away, giving a decent chance of seeing on the surface where a knapper sat to clean stone thousands of years ago, body folded over a piece of chert or night-colored obsidian, apron of flakes strewn across the ground exactly where they fell.

I reached down and planted the scraper back in the ground, one of countless stone artifacts out here. I forwent a stove or dinner in the wind. With hands in my coat pockets on a near-freezing early May evening, I waited for night, for an excuse to get in my bag and sleep.

Blustery night. No stars. The rainfly on the tent snapped and tugged. Sometime after midnight the wind must have stopped, and I could hear howls in the distance. I thought they were the low, breathy moans of extinct dire wolves, *Canis dirus*. I must have fallen asleep thinking of them. Bigger in body than modern wolves, this Pleistocene species would have sung with deeper tones: more breath, more howl. I heard one, and then another, like ragged bassoons finding each other in the dark.

This is what people would have listened to in the middle of the night, dire wolves triangulating and coming closer. This is the map they would have made, changing by the moment. Were the wolves grouping up, or just moving through? You could tell where they were, isolated individuals coming closer to one another, numbers growing.

On some nights, you'd want a cave. You'd want something at your back.

My eyes cracked open, and I listened into the dark outside my tent. The storm had stilled. Coyotes were yipping in the sage far away, howling and whimpering like puppies.

Coprolites from Paisley Caves—there are many—are kept in tape-sealed containers in a lab on campus at the University of Oregon in Eugene. These Ice Age scats have produced human proteins, and so excavators wear hazmat suits to avoid contaminating the specimens. Plastic containers in which they are stored are clear, so the dusty clods inside can be seen, not much to look at, a bit of hair and digested plant, some darkness of meat.

After a decade of work at the caves, Dr. Jenkins, a.k.a. Dr. Poop, a senior archaeologist at the University of Oregon and head of the Paisley Caves excavations, told me, "We've got horse hooves and sinew. Camel bones with flesh still attached . . . you can feel the flesh, you feel the fat on the bones. You're sitting there looking at the bone, saying this could be an animal hit on the freeway."

I visited Jenkins, a weathered, erudite field man with a gray, cropped mustache, at his lab in Eugene. He explained that he'd had a hard time gaining respect for his coprolites among colleagues. Many Paleo scientists insist he's found nothing but canid scat, oversized coyote, or possibly dire wolves that were taking shelter here. It's true that most of the feces found in these dry caves came from wolves, coyotes, bears, and big Pleistocene cats, but Jenkins stands behind his assertion that he has identified enough human proteins to represent the earliest identified human craps in America. The data are not blindingly clear. Though human proteins are present, they are liberally mixed with feces left by other animals and dusted with cave contaminants.

I asked Jenkins how he knew these were human coprolites and not the deposits of predators that had eaten humans and shat them out in a rockshelter.

Jenkins said, "Either way, it still means people were here."

"We got some American lion coprolites, which just stoked everyone," he said. "They are in Pleistocene deposits, same age as humans."

This was *Panthera atrox*, which I imagined strolling into one of these shelters, sniffing the ground as cats do, smelling something peculiar: *Hmmm. A new animal?*

At Wilson Butte Cave in nearby southern Idaho, excavations over a thirty-year period led by Ruth Gruhn, professor emerita of anthropology at the University of Alberta, came up with dates that back Jenkins's assertion. Gruhn's digs produced a basalt knife, a blade, a utilized stone flake, and two cut and butcher-marked animal bones in a layer that bottomed out at a radiocarbon date of 14,500 years ago. She found projectile points made from obsidian, which can be dated through a hydration process to between 14,600 and 13,657 years ago. These are identical to what Jenkins found at Paisley Caves, the oldest signs of what is known as the Western Stemmed Tradition, which eventually spread from here throughout the Great Basin and all along the West Coast.

The oldest known rock art in North America shows up within the same time bracket about 170 miles south of Paisley Caves. Images are etched into a gallery of dolmen-sized boulders along an evaporated lake edge in northwest Nevada, minerals in the etch marks radiocarbon dated to 14,800 years ago. The art is not like Chauvet in France, no delicate animals or handprints. The rock is porous and rough, the images entirely geometric, in patterns as complicated as stained glass: hand-sized honeycombs, geometric cross hatches, plant-like upshoots, the oldest known rock art in the Americas.

The problem with looking at the earliest evidence is that there is essentially no chance of these being the very first people. Doug Bamforth, a paleoarchaeologist from the University of Colorado, told me, "There's going to be a time before they become archaeologically visible, so the archaeology we see has to be of people who've been here a while. There was a long period when people were here that we probably have never seen. That means the folks that we do see must have known their way around."

Dennis Jenkins shares this sentiment. He emailed me, "What we have at Paisley Caves is not evidence of the first people walking over the horizon in the Summer Lake basin. One human coprolite has 9,000 *Apiacea* pollen per CC. *Apiacea* is a family of plants that generally have edible roots and tubers. The entire plant is edible but the root is particularly valuable. To exploit such plants you have to know the root is there and have the technology to get it out efficiently (digging stick and methods). In other words, its exploitation requires sophisticated knowledge about techniques, soils, and plants that had to be acquired through trial and error. This suggests to me that people were already in the region before ca. 14,000–14,525 years ago."

Jenkins's belief is backed up by an artifact recently found in a rock-shelter near Riley, Oregon, about seventy miles northeast of Paisley Caves. Below a layer of stemmed obsidian points, similar in style and age as Paisley's, is a layer of ash from a Mount St. Helens eruption firmly dated at 15,800 years old. At the bottom of this dig, twelve feet below the surface and beneath this layer of ash, archaeologists found a single artifact, a small piece of orange agate worked by human hand into a scraper. Its position shows that the scraper has to be even older than the volcano eruption, a pin in the map around 16,000 years ago.

I found the road to the caves through a few barbed-wire gates. Not enough rain had fallen in the night to slicken the route, well worn by Dennis Jenkins and his University of Oregon crews, who'd pulled out of here nine months earlier. The road led to a promontory of black rock where lava flows had piled onto each other, just after the age of the dinosaurs. They formed a ragged rock face a couple hundred yards across that once opened on a lake of glacial melt and rainwater, one of many lakes inhabiting the western third of the continent in the late Pleistocene around fifteen thousand years ago. Hard to call them caves, they look more like the twisted mouths of ogres. They are seated in the rock, a balcony view, their shelters rough and just deep enough to get back out of the weather.

I parked, put some gear on my back, and headed around the

promontory's gentle, boulder-strewn flank. Sunlight shafted between clouds, damp sage making the desert look like water. The wind coming off the dried-up lake tasted like metal. Zinc-colored dust storms were kicking up several miles out, where the lake edge has evaporated so many times in the last several thousand years it has been reduced to its elemental components: potassium, silica, sodium, sulphate. Breathing it was like licking a gun barrel.

With the glowing yellow sail of my tent set up, I sat on a black, lichen-decorated rock scanning the expanse, wishing these pegs of half-sleet would stop hitting me in the face. I had planned on staying down here for a day, getting a good eye for the topography before looking at the caves. I'd let the landscape seep into me, understanding where I am, where the people long before me must have been. The storm changed my mind, though. Why would I be sitting in the weather like a fool when a perfectly good rockshelter was nearby? I grabbed a water bottle and journal and headed for the caves.

The nearest and smallest shelter was little more than a shallow vestibule, a narrow head-high space ending a few feet inside. A big bubble in ancient lava, it was just deep enough to get out of the wind. This cave had been scoured for so long it felt hollowed out like a rind. I wondered about ghosts, if such things exist, and if the wind eventually winnows them away.

I poked my head in. "Hello?"

No one answered. No one ever does.

These caves have preserved some of the finest details, such as a piece of braided cordage or the remains of a spur-throated grasshopper, hairs and tiny barbs on its exoskeleton still intact after more than ten thousand years. Tiny, golden mandibles of Jerusalem crickets found in these caves have led archaeologists to believe people may have been eating these protein-rich insects—an easier mainstay than mammoths. From their feces, it also appears that the people relied on the root of the celery-like, umbrella-flowered *Apiacea* plant.

I entered and crouched in the back, facing out of the open mouth. Sleet spattered the toes of my boots. This, I thought, was much like the places Buck Navajo described when he pointed at the ground and told me it was where his people came from. Firelight had flickered in this cave.

Mesoamericans living in the high valleys and jungles between here and South America refer to *Chicomoztoc*, the site of the seven

Braided human cordage surrounded by insect detritus,
remnants found in Ice Age layers at Paisley Caves

caves from which the first people emerged, each cave issuing a different kind of people to populate the land. For the Chippewa, before anything else, a lone woman lived in a cave. For the Choctaw, people emerged from a mound where they crawled through a long, lightless cave out to the sunshine of the world. The Hopi, depending on the clan, came into this world through a hole in the earth, driven upward when the last world flooded.

Paisley Caves is another legendary site, this being a legend of science. There are several separate chambers, and not all have produced human feces. The shallow one I'd chosen first was an outlier, like so many others in the Upper Great Basin, turning up charcoal and Pleistocene horse hooves, layers of camel remains mixed with muskox, wolves, and coyotes. People had been found a few doors over, where dust in the rocks hold molecules of what I imagine to be an Ice Age man skinning out a badger, a woman cinching plant fibers between her teeth in the sun at the mouth of a cave.

I explored one shelter after the next, and two pigeons flapped out of the largest, where human artifacts and coprolites had been found. As the pigeons left this protection for the open sky, I ducked farther in, moving past sandbags and backfill left by Jenkins and his crew.

In the back of the cave, three perfectly proportioned eggs lay in a crib of twigs. The palm-sized nest looked like a quick attempt, no more than half an hour of pigeons scavenging packrat middens for sticks. *Columba livia* is a rapid colonizer; not originally native, it spread to nearly all of North, Central, and South America within a couple centuries of coming over on European vessels in the 1600s. The name of the pigeon's genus, *Columba*, comes from its ability to navigate, yet another moniker derived from Christopher Columbus. The name ends up on a lot of American species, including those who didn't start here. The original Latin name in the Old World was *Columba Noachi*, Noah's Dove, the one sent out from the ark in the Bible to find the first appearance of land after an Old Testament flood. The bird has found an island for itself at Paisley Caves, a small anchor in the American interior.

Native songbirds weave baskets into the grass and sage. Their speckled eggs are hidden near the ground. These birds have been here since the Ice Age. Pigeons lay eggs the color of pearl. They are softer than prairie eggs, requiring less investment in their production. This allows for a more rapid spread, exploiting niches at the edge of livability. Eight to twelve days after mating, a female *Columba* lays eggs, which hatch after eighteen days. Nesting areas that prove beneficial are kept for life, mated pairs often monogamous in order to hold the same location. If a nest is successful, it builds up. Feces and detritus pile up over time, and it eventually forms into a sort of mud pot, incorporating unhatched eggs and the mummies of dead nestlings.

I like to think of my own history as less vulgar. But it is still a history of colonization, a tale of land grabs and uprooted histories, a quickly moving people with eyes on the horizon. It might be a trait of any organism engaged in dispersal and colonization. Here they ate roots and lived in the company of horses and giant, curve-toothed panthers, stepping out of this same cave to see the same snow-topped mountains. Based on the evidence, they didn't put up their feet here for long. These caves were rarely occupied, with centuries or thousands of years between visits. By twelve thousand years ago, stays in the shelter were more frequent, while in the beginning people were few. The trickle would become a flood. The emergence would turn into an eruption.

7

A DANGEROUS EDEN

14,500 YEARS AGO

How they got to Florida no one knows. It is unclear whether the first people in the Southeast came from the Pacific coast or somehow across the Atlantic. They might have jumped to the Gulf of Mexico via Central America and traveled clockwise to the sub-tropical savanna of a Florida much larger than the one here today. Taking a different route, they could have crossed the Rockies from left to right over the continental interior, where they would have entered the courses of the Mississippi, which carried them through more than one million square miles of drainages, all leading to the Gulf. It would have been an easy net to be caught in.

Or perhaps they rose from one of the innumerable sinkholes and freshwater springs scattered across most of the state, coming up from the last world into this one.

The bread crumbs people left on that journey are too faint to call a route, but it is clear that they arrived in the Southeast by at least 14,500 years ago. Finds from the pits of the Aucilla River south of Tallahassee indicate an early widespread appearance as far away from the land bridge as people could go in North America without falling into the sea. This put them in what is now the armpit of Florida between Georgia and the Gulf, where rivers flow through birdsong, islands, and sinkholes.

Some of the old names are still on the maps in this part of the Florida Panhandle. Tate's Hell State Forest: Local legend has it that in

1875, Cebe Tate goes into the swamps after a panther with a shotgun and hunting dogs, and crawls out seven days later in a town twenty-five miles away. He is ravaged by insects, bitten by a snake, no longer has a shotgun or dogs. He tells the first person he finds, "My name is Cebe Tate and I've been through hell," and then he promptly dies.

This is not forgiving country, never has been. Florida has one of the largest concentrations of dire wolf and sabertooth cat remains in the Americas. A Southern species of short-faced bear, *Arctodus pristinus*, is well represented in the state's fossil record. Herbivores tend to be either armored in hard, hairy keratin, or armed with tusks, claws, or horns. They were in a place where one was apt to be eaten.

Near the Econfina River, about thirty miles across Apalachee Bay from Tate's Hell, two friends and I were hiking through the woods when we found a trap smeared with blood, urine, and short, coarse hairs. At first we didn't know what it was. We had to walk circles around it. Metal fence posts had been used as a frame, tied to wire fencing and greasy, battered plywood with orange baling twine, entry spring loaded.

"It's a hog trap," said one.

The other got down on his knee and looked underneath. The trap had an axle where wheels could be mounted if a captured hog needed to be transported. Two old bullet holes and the slick stains of

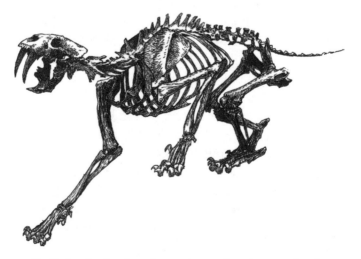

Smilodon fatalis, the sabertooth cat, abundant in Florida

blood on the wood made me think kills had happened in the box, a kicking, squalling blood spatter.

We'd tied a canoe and a kayak off along the nearby Econfina and were venturing by foot, wading through knee-deep water among knobs of cypress trees and bearded tree limbs in search of beer, a quest before sundown as we headed for a road to find a bait and tackle shop. The Econfina is a minor but navigable river flowing out of an upper Taylor County swamp and into the Gulf of Mexico, and we were paddling up its current for a couple days, farther inland, exploring the country. Spring rains elevated the water table, river coming out of its banks and filling the surrounding land. An armadillo galloped away along a spit of knobby cypress knees, like a golden ball disappearing into the woods. The hog trap was on dry land, driven down a two-track and dropped here.

"How big a pig, you think?" I asked.

"Two hundred pounds?" asked Pete as he tugged at the fence-post frame, testing its strength.

"Hogzilla!" said Alex, a tall man with a runner's body, bald, upstanding head like the top of a bowling pin, bright cheeks and a full, ruddy beard.

The two were married, both in their early thirties, and they'd taken the last name of Carr from Alex's mother. Fresh from a boot-stomping wedding at a hot spring outside Pray, Montana, they were on their honeymoon and we converged for several days, our travels coinciding in these swamps and karst rivers. Their honeymoon was a month in the Florida backcountry, navigating waterways as they crossed parts of the state in a red, tiger-striped canoe.

Pete and Alex weren't living much of anywhere you could pin down on paper. Pete, a small, energetic man of Germanic heritage, with a crop of fierce black hair, had worked the previous season as a backcountry ranger in Yosemite National Park, where he'd take two-week solos through the granite batholiths of the Sierra, checking permits, making sure no one was being eaten by bears. Alex, meanwhile, worked on a wilderness restoration crew in the same park removing trails, and before that he built trails for parks across the West and Alaska.

At the time, it was illegal for a same-sex couple to marry in Montana, where they held the wedding, so they did the paperwork in

California after waiting for an overturning of that state's proposition banning gay marriage. It was also illegal for them to marry in Florida, but neither Alex nor Pete were much into the politics of the institution. Their marriage was not an act of resistance.

They portaged by Subaru wagon, loading a canoe onto the roof of their beater car, rainbow-flag sticker in the window, "Eat Organic" bumper sticker, Colorado plates, which they used to jump between rivers, estuaries, swamps, and bayous. Alex, the tall one, had been kidding about Hogzilla. He had a perpetual smile in his eyes, as if he were holding back a joke. "Hogzilla" is the name given to the over-thousand-pound feral boars said to inhabit the South, according to likely apocryphal reports. Hunters reportedly kill them in the deepest woods, but official weights have never been taken, and photographs could have been doctored. Fish and Wildlife people say they are probably seeing eight-hundred-pound boars, the result of the inbreeding and isolation of an invasive species brought by Spaniards and mixed with every other invasive pig that came after it. That's

A mastodon molar atop a mammoth molar
show different dentition in similar habitats;
mammoths ate grass while mastodons
ate twigs, leaves, and branches.

big enough. Hogzilla, even by official definition, wouldn't fit in this particular trap anyway; he'd kick it to splinters and trot away wearing a skirt of wire fence.

Hogs were our modern-day version of mastodons, the most common large, bristling beasts of Ice Age Florida. Only one larger animal had been here, the Columbian mammoth, *Mammuthus columbi*, but Florida had many more mastodons, *Mammut americanum*. The East, especially the Southeast, was mastodon country, the animal closer in size and shape to the Asian elephant, while the Columbian mammoth was more like the African variety. Full-grown adult mastodons would have weighed six to eight tons, tusks up to seven feet long, pikes compared to the recurve of mammoth tusks, which were twice as long. The two have notably different teeth, mastodon molars being spiked, almost mountainous (*masto* referring to breast, *odon* to tooth), while mammoth molars are nearly as flat as bricks, with a reticulated grinding surface.

Skeletons and skulls show damage from mastodon-on-mastodon violence, most likely males in rut. Tusks are splintered and bones are broken in places indicative of conflicts that could have been fights to the death, bodies crashing into each other, tusks piercing hide and muscle, trees toppling in their path.

A large sliver of oak-brown bone from an Ice Age–sized animal was found at Vero Beach on Florida's Atlantic coast, with the image of either a mammoth or mastodon engraved in its surface. The drawing is simple, like a doodle, more attention paid to the position of the legs as it walked—the angle of knees in the hind legs, elbows in the fore. The head is resolutely down with tusks leading the way, trunk drawn back as if it had somewhere to be, moving swiftly through the tall grass of the temperate Florida savanna. Researchers at the Smithsonian Museum of Natural History have performed a battery of tests on patina and wear-patterns to determine the image's antiquity. Optical microscopy showed no break in coloration between the carved grooves and the surrounding material, indicating that both cut and uncut surfaces aged simultaneously, and an electron microscope revealed no recent incision or scratches from a metal tool. X-rays and backscatter imagery were thrown in for good measure, and researchers concluded "the drawing was made prior to the extinction of the mammoth, based on anatomical details, such as the relative sizes of the tusks and trunk, the high-domed head, and the long forelimbs.

This bone likely represents one of the first verified Paleo-Indian representations of a proboscidean in the Western hemisphere."

In lieu of Deep South mastodons, we had hogs. The only ones we'd seen had been trotting along the shore up the Econfina, most the size of handbags, runts and juveniles, the adults farther back, watching from shadows of palmettos and pines. The biggest adults we saw must have been eighty pounds, none quite a hundred.

We got our beer back to camp and scanned the shore for alligators before jumping into the dark, gentle waters of the Econfina. We climbed out and toweled off on woody root-bridges between cypress trees, cleaning up for dinner. No little Boy Scout stick fire, we wanted a blaze, palm fronds stacked up and consumed in the humid darkness. The plan was to drink beer and get tanked. We'd been out for a few days, sleeping among chiggers and wolf spiders, nights crackling with banana spiders, rodents, and snakes. As dusk descended, our fire went up and our belches ripped into the forest, muffled by cypress trees and their bearded, draping epiphytes, no one around to hear.

On his third beer, Pete hiccupped. His frame was lighter than either of ours. While Alex and I complained about the inefficacy of Busch beer, Pete seemed to teeter slightly.

"It's the drink of the people," he chirped, "you've got to respect the people."

When I went to the river with a headlamp, I panned the light through a haze of humidity. Away from voices and crackling flame, I stopped to listen. I had next to no idea what was out here, birdcalls I'd never heard before, a chuck-will's-widow's lonely refrain answered by another's farther away.

The Ice Age would have been louder. Micromammal remains collected from caves in the Southeast represent nearly twice the species present here today. You can extrapolate this to other fauna up the food chain, from owls to sabertooth cats. This was the meeting of boreal, steppe and subtropical taxa, one of the richest faunal regions of the Americas. Eight hundred miles south of the nearest ice. Not far from the Aucilla River, along the curve of Florida's Big Bend, southern *Arctodus* bones have been found. With pack hunters, large Ice Age bears, and sabered cats in the vicinity, a trio such as ourselves would not have sat around a fire drinking beer, backs unguarded, or wandered off alone to retrieve gear from a boat. Florida was mostly savanna at the time, drier country, an American Kenya, only with animals that would have trampled and shredded those now living in Africa.

The megafauna that humans encountered would have been naive about humans at first, less sensitive to cues signifying an unfamiliar predator. This is a known phenomenon. If carnivores extirpated from a region for half a century or more return, they have a much easier time taking down prey. A study on bears in Scandinavia and wolves in Yellowstone found that in both cases, moose were killed in disproportionately high numbers when predators returned. This supports the so-called "Blitzkrieg Hypothesis," in which humans laid waste to American megafauna by taking advantage of their innocence. Before this hypothesis is set in stone, however, researchers have also found that in both Scandinavia and Yellowstone, within one generation, the moose had adapted and the numbers of kills decreased significantly. Naïveté goes out the window fast. The megafauna of Florida or anywhere else would have quickly figured out what humans were capable of.

A 2017 report from a sinkhole along the Aucilla put humans here firmly by 14,500 years ago, with earlier, looser dates of roughly 15,000 years old coming from extensive field work in the '90s. These sinkholes acted as traps for bones and artifacts. A Pleistocene turtle was found in one with a wooden stake driven under its carapace, the stake used to pry off the breastplate of the animal to get at the meat. Ivory tools have been recovered in the depths, and spearpoints made from mastodon or mammoth tusks, barbed fishhooks made of deer bone, and bone pins that would have been used for clothing or shelters. Underwater crews have gone in with scuba gear and air hoses, digging and vacuuming under the beam of a thousand-watt spotlight in two-person teams, one holding the light, the other excavating. Organic materials survived in the anaerobic depths, according to the site publication, including "diverse plant remains, collagen-rich bone, and brain and other animal soft tissues." Parts of a mastodon were found intact at the bottom of an Aucilla sinkhole; when its stomach was dissected, it revealed chewed twigs, shrubs, grass, and leaves—woody material ground up by spiked teeth.

Sea levels were significantly lower at the time, so twice the modern-day surface area of Florida was exposed. What is now the Gulf of Mexico was another couple hundred miles of savanna. Water tables were lower, and the watery sinkholes were dry caves. Within them is a record of an early and long-lasting appearance of humans. The oldest domestic dogs in North America turned up in these holes, their bones coming out of layers dating back fifteen thousand years. These oldest dogs come from field-work research in the 1990s, the results yet to be verified by twenty-first-century techniques, but there is enough evidence to say that Ice Age people had dogs on this far side of the continent. Domestic dogs, *Canis familiaris*, may have been the only way humans could have come this far, canine support acting as camp sentries, barking at nocturnal animals and assisting in the hunt, becoming as one researcher, Stuart Fiedel, put it, "man's best friend, mammoth's worst enemy."

DNA reconstruction from prehistoric *C. familiaris* remains in America indicates that the animals derived from a common Asian population. They were not domesticated from American wolves. They came with the people.

In his work on American dogs, Fiedel has written that these introduced animals may have been a factor in the extinction of some eighty genera of megafauna. Fiedel postulates that dogs allowed humans to enter hostile territory by bringing to bay giant herbivores in the hunt. More meat was required to feed the pack, and that would have steadily upped the ante, as people gathered more protein than they needed for themselves. Dogs would have guarded camps, fending off danger, and they would have "harried native carnivores with which humans were in direct competition." The earliest dogs have been hard to find on this side of the world, but between fifteen-thousand-year-old evidence from the Aucilla River in Florida and eleven-thousand-year-old dog teeth and jaw fragments from Cueva Fell near the southern tip of South America, it's a safe bet that they accompanied the first arrivals.

Rivers seemed to arrive and disappear without notice or warning. The Econfina would spread into a swamp where we hunted for any sign of a current. The Aucilla, which we floated in a couple different segments, was sewn through the karst platform of the Panhandle, rising through skylights, flowing a mile or a few, then dropping into what are called *sucks*, sinkholes that carry water underground. You walk through a lowland forest of palmettos and loblolly pines with a kayak on your shoulder, a canoe on your back, looking for the next place where the river rises from underground. The Wacissa was the bright one, the diamond among local rivers. It flowed clean as museum glass, while the Aucilla and Econfina were brown like strong tea, darkened by vegetation in the swamps they passed through.

Along the Aucilla, we came to a limestone hole where the river vanished, driftwood and debris slowly rotating on its surface. The sink was big enough to swallow a school bus, and branches of dead and fallen trees disappeared in its depths. Roots, vines, and palms leaned in around it. Alex, Pete, and I looked in, too.

With the agility of a squirrel, Pete climbed down vines and roots that hung above the water. River sandals tucked away, clothes left on shore, he stood barefoot and naked in the curve of a palm trunk that leaned over this keyhole in the forest floor where the earth inhaled a river. He dove and his body dropped underwater like a missile.

For a moment he was gone, only a shade of himself. Alex and I waited in the trees and underbrush overhead to see what would happen, thinking he might come up struggling, the sink drawing him down.

Pete kicked upward and broke through with a gasp.

"Come on!" he called.

Alex and I stripped and cannonballed into the drink. Bubbles rose from our noses, daylight feathering into the tannic water. We unfolded our bodies and kicked upward, breaking the surface.

Though this particular hole hadn't been excavated by paleoscientists, others in the area had. An assisting archaeologist named Grayal Farr worked a site on the Aucilla known as Sloth Hole. He described the excavation to me as "thirty feet down in strong coffee." Here he found a foot-long sharpened point made from mammoth or mastodon ivory that had either fallen into a crack or been deliberately placed for later retrieval, before the space filled with water. Farr told me that on the day in question, he was the digger and his partner held the light. As he pulled out the long weapon, Farr heard his partner shouting excitedly into the water over his regulator, *foreshaft, foreshaft!* Meaning that it wasn't just some bone that had drifted into a hole. This was a lance-like killing tool long enough to fly through the ribs of a mastodon or giant bison. Under the spotlight in dark water, the weapon glowed like gold. "It's one of those hair-raising moments archaeologists have," he told me, "maybe the reason archaeologists are what they are."

Looking down at my feet, which all but disappeared in the dark, I floated in a slow circle, a history of tusks and spearpoints dropping beneath me. I imagined lightless outcrops littered with fluted stone projectiles and mastodon bones from when sea levels were lower, water tables down across Florida. It felt as if I were floating over a legend, the hole out of which people traveled from the last world into this one. Its current urged us gently down, tugging on our feet. The water cleaned us, carried down flecks of skin and detritus from our travels, adding our record to the layers below.

From a Siberian story recorded from the Mansi language: "The mammoth lives in a deep whirlpool: it has the shape of a fish, elk, bear, or horse. Animals and fish, when they become old and are about to die, fall into the whirlpool and are transformed into mammoths by the shaman of the Sky God."

Archaeologists of the future might remove layers of enamel from your teeth and detect isotopes that reveal where you were born, where you lived most of your life, and where you likely died. Your name may not survive you, but the geology you lived in will. Every place has a unique isotopic signature that writes itself into us every time we eat an apple, every time we breathe, or drink the water.

Mastodon teeth found in the Aucilla River have proven this, picking up signatures from dirt and plants digested and processed along the way. The isotopes of Florida Panhandle geology found in their teeth are overlaid with those of lower Appalachia, Georgia, possibly Tennessee, and then back to Florida. These isotopic pages show long annual migrations, much like those of modern African elephants. Mastodons traveled south out of higher conifers to graze the grasses of Florida, where they settled during winters, and turned around every spring for higher Appalachia.

Humans would have entered this maze of animal migrations. Like a dye added to swirling currents, our species would have been picked up and carried onward, propelled by dire wolf packs, cycles of mas-todons, and the arcs of rivers. They didn't land in Florida as if by parachute; instead, they arrived within the steady push-and-pull of much larger forces.

The people who first appeared in the Southeast are not clearly Beringian or from the West Coast. In fact, the earliest weapons in Florida suggest they may have been a third group, or at least came from a clearly different technology. Their weapons are not from the Western Stemmed Tradition; they are thinned at the base with a flat or concave butt rather than a tail, a unique appearance. Where this method originated is a burning question among archaeologists, answers far from resolved.

Antecedents for weapons that arose in Florida have not been found in Beringia, Asia, or anywhere along that route. Rather, this distinc-tive stone tool technology may have originated in northern Spain and France, first identified at Solutré-Pouilly in the Bourgogne region of eastern France twenty-one thousand to seventeen thousand years ago, about the same time similar technology appeared on the East Coast of North America. The cultural group in Europe is known as Solutrean, and it manufactured flattened stone artifacts, some basally thinned, bottoms thinner than the tops. This leaf-shaped style, often reduced evenly from both sides, appears on the Iberian Peninsula and along the East Coast around the last glacial maximum, when an ice bridge spanned from the British Isles to the bottom of New York.

The poster artifact for Solutreans in America is a blade picked up by a scallop dredge from the bottom of Chesapeake Bay. Matched to a maximum twenty-three-thousand-year-old radiocarbon date from mastodon remains with which it was found, this is potentially the oldest human object in the Americas. The long, thinly formed tool is made of gray rhyolite quarried from Pennsylvania bedrock, a local source fashioned in what has been called a Solutrean style.

Most archaeologists look down their noses at isolated artifacts, say-ing that an extraordinary claim requires extraordinary evidence. An Ice Age arrival from Europe is no small claim.

I visited the Paleoindian Collection at the Smithsonian National Museum of Natural History to see mounting Solutrean evidence, not just one artifact but many. Dennis Stanford, head of the museum's Archaeology Division and director of the Paleoindian/Paleoecology Program, has compiled a lab full of casts and originals from across

Paleolithic North America, a trove of blades, points, scrapers, and adzes found from Alaska to Florida, Nova Scotia to Baja California. He opened drawers in the collection, showing off casts of recent in situ finds from Maryland and Virginia. These Eastern Seaboard artifacts come from layers of earth between seventeen thousand and nineteen thousand years old, bifacial points finely worked from percussion and pinpoint pressure flaking, some of the stones basally thinned. He showed off their features as if selling jewelry. Shapes of blades were leaf-like, points shoulder-notched on one side, like what was being made in Spain and France during the Solutrean age, about the same time Dyuktai tradition was entering Alaska and Yukon from Siberia. Human evolution may have mounted into a convergence on this side of the planet, technology and the people carrying it coming in from all sides.

White supremacists have run with this Solutrean hypothesis, claiming Europeans were first, giving American racism a history it never had. Keep in mind these were *Paleolithic Europeans*, not *historic Europeans*. They were not "white." Research presented at the eighty-fourth annual meeting of the American Association of Physical Anthropologists in 2015 offered robust genetic evidence that paler skin came from natural selection around eight thousand years ago. Earlier genes suggest darker skin in Europe during the Ice Age. It was a time when race was less noticeable, when the world was more unified. Whoever they were, these new arrivals would not have been planting a flag here. They were probably lost, blown adrift, following birds and seals, and wondering where home was, same as anyone else who arrived on this continent during the Ice Age, or just about any age. One archaeologist told me that if Solutreans did show up, people would have already been here to greet them.

Stanford believes that Solutreans were the original colonizers, that the founding American population came not from Siberia but Iberia. A New Mexico–trained field archaeologist, Stanford is a sturdy bear of a man, strong hands, beard mostly gray. When I asked about genetic research that flies in the face of his hypothesis, he responded, "Bullshit."

With a slight smile under the curl of his mustache, he gave off a dual message: *Bullshit, but what do you expect when going against the archaeological establishment?* He is defying an entrenched discipline where European colonists are seen strictly as latecomers, products of Christopher Columbus, not Columbian mammoth hunters. Stanford held up two fingers and said, "Two individuals. Are you going to base the genetic history of an entire continent on two individuals? Who knows what genes you'll get when people are screwing their way across the country."

He was referring to the two oldest human remains with archivable DNA, Naia the Yucatan girl and the Anzick Boy from Montana, both about thirteen thousand years old. Neither have European genes, at least not enough to say they carried any Solutrean stock. The only European signature in their ancestry is haplogroup X, a controversial genetic signal shared by a small percentage of both European and Native American genomes. This haplogroup comes from about thirty thousand years ago, probably early Eurasians around the Ural Mountains or on the Russian Plain. This appears to be the last time a people said hello and goodbye to each other; one group drifted to the Americas, the other to Europe. The genes of Naia and Anzick Boy have only the faintest X, possibly a memory of that goodbye, but not Solutrean.

Stanford argues that another people were here besides Naia and the Anzick Boy. He showed me tools recovered from what he believes was an East Coast Ice Age boat-building site near the mouth of the Delaware. Sharpened stones and cobbles were mixed with pieces of big-game hunting weapons, diagnostically Paleolithic. He identified the tools as having been heated for pressing hot pitch into seams, the kind of work you'd find among boat builders. This is how he believes Solutreans crossed the Atlantic, ice hunters plying the floes and icebergs of the Iberian Peninsula, following what would have been enormous migrations of Pleistocene auks. This hunting lifeway is still practiced among modern Northern peoples, though on a smaller, more local scale.

Stanford explained how the cobbles had been placed in a fire, one end of each turned yellow from being heated and oxidized. He picked up one and pressed it into the air, holding the rock by an imaginary leather sheath, an oven mitt, mashing hot pitch into the seams of a skin boat that in his mind's eye was turned upside-down on a Delaware beach. That was where he'd collected these cobbles, along with the stone hammers and anvils required for working big pieces of wood.

"Tell me they weren't using boats," he said.

When Stanford showed me a drawer full of toolstone and weapons from Parson's Island, on the Maryland coast of Chesapeake Bay, I imagined their original context: a broad valley of pine barrens growing from poorly drained glacial till instead of a bay full of water. Parson's Island was a high point, a hilltop, with no significant elevation within a hundred miles, its coast petering into sloughs and tidal marshes. Any Solutreans coming by boat would have found themselves very far from home. The icy, crashing shores of Spain and France were long behind them, perhaps out of reach forever. They could have traveled seasons before making landfall, surviving in boats and on drifting ice floes. If Ernest Shackleton and his crew of twenty-eight could survive seven months after their wooden ship sank, trapping them on Antarctic ice in the winter of 1915–1916, certainly Indigenous sea hunters—raised in Ice Age conditions—could have endured ice and water crossings for much longer. They would have finally beached along sprawling river deltas on the exposed continental shelf of the Eastern Seaboard, where big, unfamiliar proboscideans would have gazed at them, strangers in an unknown land.

When I saw the Parson's Island artifacts, I thought they didn't look American. Then I wondered: What *does* look American? Being from this continent has never been one thing. Everyone here came from someplace else, even if from underground. The stories are all about arrival.

The Parson's Island artifacts appear to be Solutrean and in stratigraphic context are twenty-one thousand years old. For more than a century, tools and weapons like this have been turning up in Maryland, Delaware, and Virginia, catalogued in the Smithsonian Museum as "Solutrean artifacts." They were originally believed to be mementos carried by eighteenth-century European colonists, Paleolithic tools as souvenirs to remind new Americans of home. This explanation for European Ice Age artifacts on this side of the Atlantic, however, seems unconvincingly vague, similar to imagining the eleven rocks at Snowmass rafting in on ice and sinking down to land, serendipitously, right on top of a mammoth carcass. Solutrean-like artifacts, like those from Parson's Island, are being found in Paleolithic layers, meaning their age is becoming hard to argue with. Stanford believes this was the start of Clovis technology, which would soon sweep the continent. Evidence of the connection has been slow to

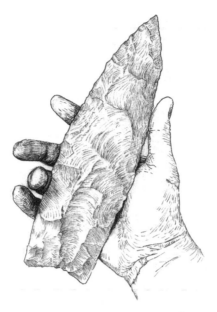

One of the largest Clovis points from
a cache found near East Wenatchee,
Washington, cast and curated at the
Smithsonian's Paleoindian Collection

take in the archaeological community. Refuting similarities between Solutrean artifacts and those found in the Americas, archaeologists have explained their similarities as "the well-known phenomenon of technological convergence or parallelism."

Whatever they were, Clovis points that may or may not have arisen from the Solutrean tradition were unique and spread rapidly. Some were larger than almost any stone weapon technology in the world. Bruce Bradley, an American lithics expert teaching Stone Age technologies at University of Exeter in the United Kingdom, calls this kind of projectile "symbolic and beyond function." He told me he's seen Clovis points two feet long. "You can't use that for anything," Bradley said. "They're out there showing off, it becomes part of the symbolic aspect of their culture."

Giving Stanford and his colleagues, like Bradley, the benefit of the doubt, Solutreans could have kicked off this new American technology. But they didn't have to physically colonize the place to do it.

Like the Vikings who came in historic times, they may have landed and died out, leaving only artifacts, signs of themselves that would have endured for thousands of years. In this scenario, I imagine Solutrean artifacts being found by people crossing the continent from the west, people coming from the East Asian genome of Naia and Anzick Boy. Somewhere on the Eastern Seaboard, these people of Asian origin may have stepped through birch groves and tundra, crouching at a ring of lichen-covered stones, startled to find signs of an earlier people where they thought no one had been before. Digging through duff and lichens, someone may have uncovered a Solutrean blade. The tool would have been unfamiliar, pressure-flaked and reduced from both sides, remarkably long and robust, yet thin and easy to carry. Those who left it were not barbarians. They would have been another people like the Beringians or those from the Pacific Rim, versed in fine tool-making, a tradition brought from far away.

If you'd found one of these artifacts, you might have worried the flaking with your thumb, as one does with an old point or stone blade picked off the ground. You'd study it, maybe huddle over it with others to understand how it was made, wondering who might have fashioned it. You might test the technique yourself, seeing whether you could get a similar result out of a piece of North Carolina chert or an agatized coral from Florida. Sharp and strong enough to hamstring a mastodon, this new projectile would have had an attractive sheen. Imitations may have spread, new traditions formed, and the Eastern Seaboard down to Florida would have become the home of a mixed and revitalized technology.

Around 13,500 years ago, a new characteristic appeared on this old, potentially Solutrean design, making it distinctly American. People began to flute their points up the middle, notching both faces of the projectile, as if adding a faux Pacific Rim stem to the shape of an Iberian leaf-shaped point, marking it not Solutrean, but Clovis.

The jury is still out on Solutreans in America. While the hypothesis of an Atlantic arrival is appearing in more textbooks and on *National Geographic* maps, scientists are not in agreement. It's a lot to swallow. If Solutreans did show up on these shores, they might have stayed long enough to leave artifacts but nothing else. If they hadn't arrived, we'd never know.

Alex, Pete, and I hired a car shuttle from a man named Jim who lived in a stick frame house back in the woods near the clear, spring-laden head of the Wacissa. In a rusted Chevy pickup, Jim dropped Alex and Pete at the river and drove back with me to move our Subaru to where we could hitch out and pick it up from the next river. We'd spent the night on Jim's land, setting up tents in spiny locust trees and lowland pines. Waving a can of beer, he'd told us that the hundreds of emerald-green pinpricks of light we'd been seeing in our headlamps were the eyes of wolf spiders. They naturally oriented to our lights, making us feel as if they were all staring at us, which they were. His driveway was made of hundreds of blue-and-white Busch cans flattened by his tires. Pete was right, it was the drink of the people.

Jim had a sinkhole on his property, a modest, grassy maw full of water. Deep inside was what he thought was the tusk of a mastodon or maybe a mammoth. He was careful to mention both, knowing that both animals had lived here. At night, he'd shine a spotlight rigged to a car battery into the hole and he could see something gleaming down there. He said he notified the university—not remembering which university—which sent two divers, who came back up saying it was just a tree branch stripped of bark, the wood preserved in the anaerobic depths of clear, underground water. Jim wasn't convinced.

Jim was wild-haired and kind. The dashboard of his pickup held a small library of field guides to birds and plants. I followed him in the Subaru out on muddy roads through clear-cut plantations and dense woods, then he brought me back in his pickup to meet Pete and Alex on the headwaters of the Wacissa, where a black cottonmouth hung like a rope in the branches above the water. He talked about where the rivers go, the ones that disappear, the ones that merge and separate. He talked about plantations and muddy roads. We drove through a small African American town, not a white face to be seen, and he told me the town had been here since anyone can remember, its founding going back to emancipation, probably older. This was thirty miles into the backwoods southeast of Tallahassee, one of the last Confederate capitals to fall during the Civil War, and it must have felt like the American Congo in 1863, the year of the Emancipation Proclamation.

This has always been complicated country for arrivals, an Eden, but a dangerous one. Indian wars, slavery, and colonial atrocities are embedded in its history. The first African American town was established in Florida several decades before the Revolutionary War. Its fort was manned by fugitive slaves, an all-black militia led by Francisco Menendez, the first African American military leader who fought for the Spanish in the South against the British encroaching from the North. Before him, the first recorded African in the New World was Esteban the Moor, a Moroccan slave acquired in Portugal by a Spanish nobleman, which landed him a spot on the 1527 Narvaes expedition to colonize Florida. After a shipwreck near Tampa Bay, he was one of a handful to survive and make it three thousand miles in nine years to the nearest Spanish outpost in what is now Mexico.

We dipped paddles, starting for the clear current of the Wacissa, when Jim asked from shore, "You know how to get to the Gulf, right?"

"We head downstream," said Alex, turning back from the bow of their canoe, the river beneath him as transparent as glass. "All rivers lead to the sea," he said.

Jim hiked up a belt loop, not sure how to explain his rivers to people entirely unfamiliar with local hydrology. He shouted last-minute directions: *Down past Goose Pasture on the right, turn into the grass, it doesn't look like the way, but it is.*

Snakes in trees, who thought of that? We saw one as big around as the thick end of a baseball bat, armored as if for combat, ridges down its back, sliding from a branch into the water with the tip of its tail last to go. It was a cottonmouth.

A friend who grew up in Louisiana told me of being in a boat with his dad when a big snake dropped in from above. When it hit the floor, it sprang like a whip. He remembered climbing on the gunnels while his dad grabbed a shotgun. The snake was a cottonmouth, which delivers a hemotoxin, a blood killer. The pain of its bite has been described by snake experts as almost unbearable. It is frequently fatal, and those who don't die often need skin grafts to fill necrotic craters left in the flesh around the puncture wounds. With his shotgun, my friend's dad started blowing holes in the bottom of the boat till he killed the snake. Better to plug up holes later and face alligators than to have a live cottonmouth in your boat. Apocryphal or not, I took his story to heart. Don't let a snake drop in your boat. It was one of the last things Jim told me, don't get a snake in your kayak.

We couldn't go a minute without something alighting on us or scrambling in the brush, butterflies, squirrels, and little black coots grabbling and screeching among reeds. Egret, ibis, osprey, and heron burst out of the foliage while helmets of turtles dropped from branches into the river ahead of us. Jim had reminded us that this is the part of Florida where the ivory-billed woodpecker, thought to be extinct since the 1940s, was—allegedly—spotted by a team of ornithologists in 2006. We might get lucky, he said.

A log floated nearby, stilled in a slight eddy behind a snag of branches. It was a fallen, waterlogged tree with only a spine sticking up. I paddled closer until I could see eyes like clear glass marbles. It was not a log. Without a stir, the alligator was gone.

Before this trip, I'd never seen an alligator in the wild. Here they popped up at every river bend, none longer than five feet, more fascination than concern. The most recent alligator death in the US that I knew of was a twenty-three-year-old Texas man at a bayou marina on the Louisiana border. It was a summer night just past midnight on a near full moon, and he ignored posted warnings about the twelve-foot-long alligator seen in the area. The last thing he was heard saying was, "Fuck that alligator." Both he and his girlfriend jumped in, and a large shape moved out from under the dock. The girlfriend made it out, but the young man did not. She climbed onto the dock while he screamed and thrashed. The screaming stopped, water stilled. Moments later, under the moonlight, his body rose to the surface, motionless. The alligator emerged and dragged it back under, while the young man's girlfriend watched in shock.

Fuck that alligator, or any other large, predatory animal, might not have been a Paleolithic sentiment. At least not one often talked about. There were so many other dangers on the frontier of Florida. Pleistocene alligators were the same species, *Alligator mississippiensis*, and grew up to twenty-two feet long, almost twice the size of the one that killed the young Texas man. For anyone coming from glaciers and giant Ice Age rivers out of the continental interior, twenty-two-foot-long reptile predators would have been a shock.

I thought about dire wolves as I walked through pines and palms, my kayak tied back along the Wacissa. Pete and Alex had pulled in their canoe downstream and tied off, exploring by foot to see where we were. Looking through the forest, I thought I wouldn't see wolves until it was too late. It was hard to spot anything through vines and palmettos, pine trunks standing among them as straight as telephone poles. I didn't know the plants here, felt as if the ground were booby-trapped. Which plants were poisonous, which weren't? Poison ivy, oak, and sumac, stinging nettle, a stinging perennial spurge called bull-nettle, and tread-softly, few of which I knew from home. I couldn't tell which to touch and which to push away with a stick. Jim would have known, but we hadn't asked.

Pete grabbed hold of a rope-girthed vine hanging down a pine trunk and clambered his way up. Twenty feet above us, he said, the view wasn't any different. It was just green and green.

I waded into fans of palmettos, waving away mosquitoes, thinking the wolves would have been watching from the shadows, their eyes between trunks and palmettos, seeing the first strange people. We spread out from each other as far as we wanted. I could barely see the other two, and then I couldn't. Easy to get lost, flickering sun through the canopy. I scanned the woods around me. I could see the way back to our boats—a big walnut tree, a press of grass where I'd walked, a broken flower stem. The map, with its vanishing and merging rivers, was all but useless. We used our bodies instead, trying the place the way humans have done from the start.

Animals would have acted as maps, indicators of whatever lay ahead. Sue Ware, the paleontologist who determined that dire wolves lived in packs, believes they would have assisted human entry by showing how to live in the place. The wolves, the animals most like us in their ways, would have been especially helpful.

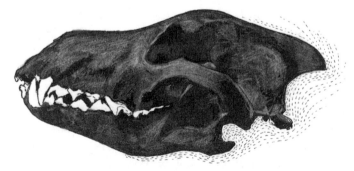

Skull of the dire wolf, *Canis dirus*

"There would have been a period of observation," Ware said. "Humans would have watched the animals to see how they navigated the landscape, how they found prey, how they avoided being killed. These guys aren't going to run pell-mell with spears into a herd in a place they don't know with animals they don't know. Any predator will arrive and assess what they're up against."

The seasonal migrations of mastodons would have told people of farther lands, wintering and summering grounds they'd not seen with their own eyes. Where to find fresh water and meat, where to find shelter—these are the things humans needed to know.

Animals may have offered more than routes. Some would have lit a fire under people's feet, keeping them moving. Wolf packs do not suffer competition. They are known for aggravated attacks on other canids, killing off coyotes, dogs, or other groups of wolves. They may have given people the same welcome. At first, wolves would have regarded our kind as exotic, confusing, the range of a thrown weapon not within their memories. Domestic dogs, bred from Eurasian wolves, must have been bewildering. Wolves would have understood trans-species symbiosis, following ravens to weakened or injured prey, but not like this, traveling at another species' side as if the two had become one animal. For the smart wolves, the novelty would have worn off quickly. As odd as humans seemed, they were ultimately competitors, going after the same prey in the same way, pack hunters, just another kind of dog.

"Introducing humans with fire and stone tools changes the playing field," Ware said. "When the wolves find out we've got weapons, the dance begins."

The common bison living in Florida, and across nearly all of North America in the late Pleistocene, was *Bison antiquus*, its horn corns slightly shorter than *B. latifrons*, and curved at the tips like those of modern bison. A humpbacked giant, the spinal process between *antiquus* shoulders lifted like a hatchet at a maximum of seven and a half feet tall, with a body fifteen feet long.

In 1981, on the clear bottom of the Wacissa where the broad river bends like an elbow, the skull of such a Pleistocene bison was found by three locals. When they pulled the exposed skull up from the riverbed, they had in their hands the horned block of a mature female that would have weighed around eighteen hundred pounds. It had a piece of stone projectile sticking from the center of its forehead.

The point, a Suwannee chert used frequently for a local toolstone, penetrated the sinus cavity but did not go any farther. It entered the bone when it was fresh, when the animal was alive, or very recently dead. X-rays show that the projectile did not reach the brain. If it had been bound to a shaft with sinew and mastic, a good spear-throw into this bison would have been from fifteen to thirty feet away. If the weapon had been an atlatl dart, which is proving to be more likely the case for late Pleistocene projectiles, the thrower may have had another eighty feet of range, a distance that could have been closed in seconds by a charging *antiquus*. The closest modern equivalent of this encounter would be the Cape buffalo, about half the weight of *antiquus*, known to have killed more modern hunters than any other African animal. Enormous and primed for attack in a country of aggressive predators, Cape buffalo have been widely documented killing lions, surrounding them, whipping them into the air, pummeling them into the ground with horns and hooves. The higher the stakes, the more hostile the animals, putting the Ice Age into a category well beyond modern Africa.

This encounter in Florida might have been head-on, the stone tip thrown by the hunter as a last resort, into the forehead, where it snapped, followed by the impact of a skull the size of a rain barrel against a human body. The hunter would have been rag-dolled into the air, caught on a horn tip, and thrown.

Whether the person who last wielded this piece of Suwanee chert survived the encounter is unknowable; there may have been time for a hard stab of a blade, pounding a hunting knife into the thick hide of its head after it fell in the hunt. What is known is that it happened. Giant Ice Age bison and human made clear and visceral contact on what is now the Wacissa River.

We camped at the river's wide elbow where the *antiquus* skull was found. It was a known site, a campground Jim had told us about, a half-circle of moldering camper shells, a fifth wheel, and a canvas tent of 1970s vintage. As far as I could tell, the site's occupants were white guys maybe thirty to sixty-five years old, no women. They seemed like wanderers who knew each other, jeans and denim coats, throwing split logs into a fire and shooting the shit. The site was cut from the forest, camping pads cleared around a half-crumbling concrete loading ramp into the river. After a century of disuse, it would be barely visible, perhaps some of the concrete around the fire pits still visible in the piney wood duff. In a thousand years, a detailed surface survey would be required to define the camp's presence. Fourteen-thousand, five hundred years from now, through hurricanes and sea-level fluctuations, the river whipping through centuries like a snake, it's as good as gone, lost from the archaeological record.

After the first glance at us, the guys around the fire didn't look over again. We'd be out of here in the morning.

While our backpacking stove hissed, a gray-bearded gnome of a man in suspenders and a baseball cap strolled over from the RV. He said his name was Ed. He tucked his hands in his back pockets and in a welcoming drawl said, "You boys enjoying your trip?"

It took a moment to put his words together. I knew the drawl of southern New Mexico, but this was a different pattern, the long, slow diversification that languages go through before they split, before new families appear and old tongues are forgotten. Give it a thousand years and we might have no words in common.

But we understood each other perfectly well, more through faces and smiles than words. Effusive, we told him the river corridor was much wilder than we'd expected. No houses, clear cuts, or power lines, it was beautiful. I'd never been in Spanish moss and cypress, it was like *The Pirates of the Caribbean*. Ed laughed. He knew where we were going. We didn't have big knives or guns, armed with just what we needed for traveling through. He warned us that channels begin to divide from here. The river braids and then starts falling into holes. He was born and raised around here, he said, came from a town whose name I hadn't known, enjoying his retirement alone. The RV didn't look as if it had moved in months. When he saw we didn't have any wood and our blackened fire pit grill installation was empty, he said, "Y'all need a fire." He told us he'd bring over two or three pieces of good, dry oak.

We didn't have any beer to offer, really nothing at all. We were wanderers, packed tightly. The kindness of strangers was all we had. Ed seemed to be the lookout, the ambassador for this semicircle in the woods. The next morning, he'd help us arrange another shuttle for our Subaru from a guy at the campground named Bobby, who would spit a string of chew on the ground and wipe the back of his hand across his mouth just as I was reaching out to give him a shake and introduce myself. Friendly enough, but don't touch.

Ed came back with an armload of split oak, a hatchet, and a bundle of tacky, amber-colored pieces of wood. As he knelt and splintered the ends of the sticks with his hatchet, he said we should know how it's really done around here. He said this is called *sapwood*. "You're gonna get yourself some of this," he said. "They used to make gunpowder out of it. Now it's synthetic stuff, man-made junk. This is the real thing."

Ed flicked a lighter under the splinters and a flame took without hesitation. He put the sapwood under the split wood and had a fire going with just a couple quick puffs of breath.

We told him where we were from, and Ed smiled, nodded, asked questions. When Alex and Pete mentioned being married and this being their honeymoon, I cringed a little. I'd heard rumors of the Deep South, the underside of Georgia. Panhandle highways were dotted with big white tents, belonging to a religion I knew next to nothing of, something I'd heard about snake-handling and speaking in tongues.

Ed said he was glad for them. He said that everyone should be able to love whomever they want.

Maybe he couldn't care less about sexual orientation, skin color, or what kinds of weapons we did or did not carry. Maybe he would have come down to the river to greet anyone who showed up, it was enough just to see we were human.

Hands and feet, fire and tools; these likely would have been better to see than fangs or tusks. A moment of watching, and then a smile, a smile in return, a decent gesture in the middle of nowhere. I asked about the Ice Age, and Ed mentioned a scientist who stopped in and told him about looking into the clear water of some braid or river bend on the Wacissa and seeing the articulated bones of, what was it, a bison, maybe a mastodon? It seemed like anyone we talked to around here had something to say about the Ice Age, a comment at the tackle shop, Jim in his yard with sinkholes, Ed on a bend of the Wacissa looking into the water, hoping to see something.

We told him we were looking for a place called the Slave Canal, which Jim had told us about, which would eventually get us to the Gulf.

"Drunk Jim?" Ed asked, nodding upstream.

Drunk Jim. Now we were networking.

Ed said it's the only way to get from the Wacissa to the Aucilla without getting lost in a bunch of sinkholes and bogs.

"Once you're in the Slave Canal, whenever you have a choice, stay right," he said.

"How easy is it to get lost in there?" Pete asked.

Ed tipped back his baseball cap and said, "Now, how you gonna get lost? You go through the woods, you end up comin' out over yonder."

Yonder? I wondered. How far is that? *Where* is that? What would Cebe Tate say?

In the Ice Age, yonder could have been twenty-three hundred miles across the continent, over the Western mountains to the coast where people were moving by boat, eating salmon and mastodon. It could have been the Gault site in Texas up Buttermilk Creek, an outpost of early Clovis, where there were bison hunters and butchers of mammoths and deer. Wherever we came from, Ed seemed sure we'd find our way back.

Not wanting to overstay his welcome, he stood from a crouch. With our fire up and going, he turned to leave. As he straightened his ball cap, he looked back and said, "You go and tell people about this place, tell them how beautiful it is."

Airboats appeared after dark. The first truck trailer backed into the water, headlights of another truck illuminating the work as a crank was let down and gear was hauled onto the boat. To get here, they'd driven through half an hour of mud ruts and pools, the only way to reach this ramp and river bend.

Through the unzipped top of my tent door, the ramp looked like a paramilitary expedition with beer. I saw a rifle being handed up, a .22, better for killing an alligator than a larger caliber that would damage an otherwise valuable skull. Their flat-bottomed jon boat was made of aluminum with a wire cage around the fan, something that could have been welded together in a garage. It cranked up like a tractor. Fan blades coughed and picked up speed. Spotlights rigged to the cage flicked on. The forest canopy lit up as the boat moved into the water.

Another came after that, and then another. They were all heading for what Ed had described as a dead end, a slough of bogs and reptiles in the lower Wacissa, not a good place for a canoe and kayak trying to reach the Gulf. Five or six went out that night, and I fell asleep to the half-truck, half-airplane sounds of engines throttling up and down, pilots sitting high, navigating the braided bottom of the Wacissa by the swivel of spotlights. By law, alligators have to be caught with a baited hook and line and brought to the side of the boat, where they are secured before being shot in the head. This prevents a free-for-all in the swamps, people taking potshots at anything that moves. It also means the reptile is fully alive and close enough to touch, ready to fight to the death.

In the fall of 2010, a woman named Mary Ellen Mara-Christian killed a thirteen-foot-long, nine-hundred-pound alligator in a South Carolina swamp. Her kill was one of the largest alligators on record, and a local news crew came to the ramp to film her. The behemoth was hoisted and dragged by a tractor, draped across a concrete slab where they set up their shot, barely able to get the full length of the alligator into the frame without the hunter looking small and far away. The hunter was visiting from Massachusetts and had a lottery tag for killing big reptiles. She wore coverall shorts and pink flip-flops. She told the camera that her hands were still shaking now, hours later. She described hooking the monster on hundred-pound test, and spending two hours reeling it to her boat before she "stuck it" with two harpoons. She said it felt like a truck underwater. Once she had it secured, she fired shots from a handgun into its head. Instead of dying, the alligator whipped around and banged into the boat, its jaws flaring. She unsheathed a knife, leaned onto the massive head and stabbed repeatedly into its spinal cord. Only then did it die.

Crouched behind the carcass, her hands flying as she breathlessly described what had happened, she said, "Your heart's just pumping and pumping, and he was enormous, I mean look at—you can see this part of him coming out of the water, his big belly, and then, then you saw his head, 'cause his head had to come out for me to shoot him."

We've been doing this for a long time. The hunt is written into our muscle memory; it has to be, as killing animals is a way that we survived. The shores of Lake Victoria in Kenya have revealed animal bones cut with stone tools two million years ago by a *Homo* species, possibly *erectus*. Dozens of goat-size gazelles were brought to the site in what appears to be one of the earliest documented hunts, at a time when human ancestors were moving from forests onto the savanna, leaving shelter to fend for themselves with sharpened implements. Soon we'd be big-brained *Homo sapiens*, wielding bifacial tools, stemmed projectiles, and leaf-shaped lances made of stone. We would rule the world.

An Ice Age hunt must have been musky gore, dogs ripping at the air, projectiles sailing, their shafts wobbling in flight before a tapered point sank into flesh or bone, an atlatl able to deliver a Clovis point at 76 miles per hour. Killing big animals is a trait that came over with the first people. A domestic dog found buried at a Siberian site had a piece of mammoth bone inserted into its mouth, perhaps a scrap from the hunt, an offering to a trusted companion on the way to the far side of the world.

The hunt doesn't look like a museum diorama, according to George Frison, a revered professor-hunter-archaeologist of Paleo-Indian sites on the Great Plains. In his book, *Survival by Hunting*, Frison wrote that most museum displays and dioramas give "the erroneous impression that wild animals make little effort to evade human predators and that delivering lethal wounds with primitive weaponry required only that the hunter approach within a short distance of an animal, and then thrust a spear or launch a dart with an atlatl; the target animal seems to regard the entire operation with detachment, apparently embarrassed to be involved in such a ridiculous episode."

Frison is a lifelong hunter. Born in the fall of 1929 to a homesteader family near Ten Sleep, Wyoming, he would come to spend a career applying his sense of terrain and the movement of animals and humans to the ancestral ecotones of the Great Plains and Rockies. Frison surmises that topography and the way it carries animals across itself is more or less the same today as it was in the late Pleistocene. He hunted rolling swaths of prairie and mountainous terrain, where he saw a place that was relatively unchanged, and a way of being that humans have not forgotten. He dumped Asian elephant bones into a stream to see how different levels of flow would take the bones apart and scatter them downstream. By doing this, he could better understand the difference between mammoth bones collected at a kill site and those haphazardly strewn by erosion. Showing a conceit that might be specific to his own habitat, or at least his upbringing as a Western hunter, Frison wrote, "I strongly believe that only rarely, under unusual circumstances, would a reputable hunter scavenge dead animals."

Florida may have been different. With a heavy herbivore load—many paleontological sites reveal the remains of large grazers—scavengers would have been everywhere, teaching humans how it was done. Like modern wolves, dire wolves would have found caloric value in fighting over a fallen carcass instead of taking it down themselves. Meanwhile, the teratorn, *Teratornis merriami*, a condor-like carrion eater common to the Southeast in the late Pleistocene, would have come in for landings on a twelve-foot wingspan, big enough to defend a carcass from big cats, dire wolves, and southern short-faced bears.

One of the regular cats lurking in Florida was *Homotherium*, the scimitar cat, another feline that weighed up to five hundred pounds (a hundred pounds heavier than African lions). Unlike *Smilodon*—another predator common to Florida, with long, sleek puncture-blades—the scimitar cat's canines were curved and elongated, more like buck knives than scythes, serrated to both stab and saw. While *Smilodon* had to be careful with its sabers, which were prone to breaking, *Homotherium* could be more liberal with its assault, jaws opening 90 degrees for large bites, teeth positioned to peel back slabs of flesh. A *Homotherium* kill would have been more violent than an attack by sabertooth or American lions, involved more kicking and screaming, flesh pulled off in steak-sized hunks. I can understand why the first people here might have climbed into sinkholes, or at least lingered around them and left artifacts and detritus. They needed protection at their backs, shelter they could jump into, spearpoints aimed up at the circle of sky.

The scale of predation here is told by the animals that survived it. The armadillo we saw on the Econfina was a nine-banded species, ten pounds at most, the size of a football, its armor able to deflect a bobcat or a coyote. Its Ice Age relative in Florida reached up to three hundred pounds and was defensively plated, an evolutionary import from South America. Known as a *pampathere*, this giant armadillo was plated head to tail with hairy, coarse keratin armor, the thickness of its shell directly proportional to the bite force of the predators trying to get through. These were grazers, consuming mainly coarse vegetation. Another local, slightly larger grazer was the glyptodont, *Glyptotherium floridanum,* a one-ton armored mammal found from Florida to northern Mexico. Its shell was rigidly fused keratin, a hard, knobby skull plate, and a club-like tail encircled with spiny rings. Fringes around the heads of glyptodonts were sometimes lined with spikes facing outward like a collar of sharp teeth, guarding a cudgel of a head. Part mammal, part dinosaur, part *Mad Max* war machine, these glyptodonts give some idea of what was needed to survive among resident predators.

At a dig lower down in Florida, the Millennium Park site near Tampa, about 53 percent of the identifiable mammal bones belonged to armadillos, pampatheres, and glyptodonts. The abundance of her-bivores suggests available meat was everywhere. While Frison was crusading across Wyoming, killing mammoths and giant plains bison with atlatls in his imagination, these people were hunting in Flori-da's tall grass and deep tangled woods. I have wondered why the Old World saw significant cave art tens of thousands of years before art arrived in the Americas. People in the Old World had more time to settle in. Humans and the hominids before them lived near caves for hundreds of thousands of years before any painting began. Eventu-ally, they cobbled together some kind of brushes and started a little something on the wall. Not so in the Americas. Animals had been left to themselves on a continental scale for ages, with little to no warning of humankind. Instead of being a gradual advance, as the Old World experienced, here the contact happened in a snap. In the beginning, in Florida, it would have been a shock-and-awe campaign on both sides, little time or inclination for making art.

Slow, tight strokes took us around fallen trees where I nosed the tip of my kayak between root burls and leafy, spider-strewn strainers. The Slave Canal was a mess. A century of hurricanes and floods had blown it apart, turning it wild again. We had to climb over some of the fallen trees, hauling our boats behind us. Braided channels went everywhere, disappearing back into the woods where the most recent hurricane had coughed up enough debris and uprooted trees to block some of the right turns we'd been advised to take. Purple irises span-gled the shores and vines strangled their way up trees. Giant spiders, golden silk orb-weavers, locally known as banana spiders, hung in the canopy like circus performers, some nearly as big as our outspread hands.

Paddling out of the main current and into deeper backwaters, we listened to the chirp of young alligators where boats don't go, the river breaking and breaking again. I looked back into a small tree-trunked cove where little alligators slid from their mother's back and fished through the water. The mother sank to her snout, so I could only see her nostrils, eyes, and ridge of a tail. She turned to watch me pass. We made eye contact, if that's possible with a reptile. Her babies swam around her like pollywogs, little tails working the water.

The original canal walls built by slaves were still visible, rough stacks of limestone threaded with roots, half-covered with leaves under squat palms and long-leafed pines. Each rock had been broken from the ground with a pickax, pried up and stacked. I reached out and touched one as I passed, the limestone pitted and moss-bound, history stacked on history. Around 1850, slaves were sent out by cotton plantations around the head of the Wacissa to build a new channel to take cotton barges to the Gulf. The Wacissa falls into sinkholes and sloughs, no place for a barge, so another route was needed. In the heat and pit vipers, they chopped a new water route across ten miles of lowland swamp and forest with pickaxes and shovels, breaking out pieces of limestone floor and building them into canal walls.

The canal took them through a region of hundreds of Timucuan mounds, remains of prehistoric chiefdoms estimated to have reached up to two hundred thousand people, extinct by the start of the nine-teenth century, their language lost. Diggers must have encountered artifacts as they trenched across the lowlands, impossible not to see colorful Timucuan pottery and glassy rock chips in the soil. Down deeper, they would have found Archaic shell middens and piled-up encampments. Slaves might have been the first archaeologists, rec-ognizing cultural distinctiveness in different horizons of earth. The deeper they went, the larger the stone weapons would have appeared, breaking into lenses of ancient campfires, layers of earth darkened by ash and cooked bone. In the deepest levels were mastodon bones and tusks, and the skulls of sabertooth cats and American lions.

A man laced with sweat might have held up a fluted marvel, a stone projectile for taking down beasts twenty times his size. It must have conjured a scale of animal that could have only existed in legend. Turning the long and elegantly flaked weapon, wiping off its surface, the digger must have wondered what history was in this ground, what new country he'd discovered.

The largest, most refined weapons in ancient America were seen on this side of the continent first. Once you see one of them, its shape is unforgettable. A piece of stone is reduced from both sides until it is thin, light, and long. The projectile conspicuously lacks the tail of the Western Stemmed Tradition, and instead has a concave butt, shedding a small amount of weight from the back end. This technology was nuclear for its age, the height of Ice Age stonework, lightweight yet strong, able to break in such a way that it could be quickly repaired. There was a confidence in the work, knowing exactly how a stone would respond to blows from an antler or from another stone.

Modern-day professional knappers, people who replicate ancient tool work, will tell you that making one of these is an aesthetic statement, a fine skill that went beyond mere function. Archaeologists in Florida working the Aucilla River in the 1990s, and other sites from here on up the Eastern Seaboard—the Topper site along the Savannah River in Allendale County, South Carolina, and the Cactus Hill site along the Nottoway River in Virginia—have found many of these fine, early points. Topper and Cactus Hill revealed them in layers of dirt and sand dating to eighteen thousand and twenty thousand years ago, possibly echoes of a Solutrean landing. They might have needed more refined weapons in the deadly landscape of the Southeast, forged on the savannas of Florida. Smaller stemmed points and atlatl darts may not have been enough.

Around thirteen thousand years ago a new signature appears on these weapons. They become "fluted," the word coming from the fluting of columns in Greek architecture. The marks were made by single, well-placed blows that knocked a channel into the projectile. The purpose of fluting is unclear. It could have assisted attachment to the shaft, or taken weight off, enough so it could be thrown from an atlatl with remarkable accuracy. One suggestion has been that it channeled blood, giving an airspace large enough that an animal would be weakened and fall sooner than if the point were firmly embedded in flesh.

The fluting serves as a stylistic signifier, a way of expressing identity. After Pacific Rim stemmed points, this is the next cultural appearance, indicating not just people, but *another* people.

Alex bears a tattoo on the softer underskin of his left forearm. Gothic black letters spell out *Sero Sed Serio,* his Scottish clan motto, from the Battle of Ancrum Moor in AD 1545. "Late but in Earnest": the motto of Clan Kerr. The Kerrs were mercenaries that the British brought to this borderland battle against the Scots. They waited to enter the fight, and at the last moment joined the Scots. Now replenished with fresh, strong troops eager to fight, the Scots drove back the Brits. Clan Kerr turned the tide, carrying into victory their new motto, which might also be translated *Better late than never.*

Alex had recently returned to those borderlands, visiting the castle his people built. He sat beneath an aged oak tree, its heavy limbs held up by wooden supports, the same oak that stood there in 1545. Sitting under that tree with his brother, in the very place where their clan gained its motto, he felt a sense of origin he'd never known before. This was his oldest sighting of his own bloodline. He told me, "Even if you are not going to live there, even if you aren't going to return, you have that zero-latitude, zero-longitude from which to understand the world."

We each have a place and time to remember. Alex's is along the Scottish Borderlands, while many Native Americans hark back to springs and holes in the ground, or a headstone in an Alaskan forest in Tlingit territory. I have a turtle tattooed on my back as a compass, and a raven feather on my shoulder from the Grand Canyon, where I watched two ravens battle a falcon. A lone black feather floated to the ground, and I picked it up, a talisman. Raven, who stole the sun, moon, and stars. Raven, who flew across an ice sheet. These are my ways of recording my place. Maybe I've jumped ship, defecting to this continent. I am European, with the bloodlines of Wales and Germany, and I live here in North America, my oldest personal memories coming from this new ground.

I came to Florida looking for the human source on this side of the planet, the holes in the earth from which people far older than I have emerged, some of the oldest widespread archaeology in the New World. This is where the Slave Canal led us onto the broad, dark Aucilla, a river that people followed in the Ice Age, their artifacts continuing even off the land and out into the Gulf, where sites are found by sonar and excavated with vacuums. We passed beneath the concrete-pylon bridge of Highway 98, about five miles from where the Aucilla opens onto the Gulf. The bridge surface is made of metal grate, which caused a terrific roar whenever a vehicle passed overhead every minute or so. The Wacissa had turned into the Aucilla; both rivers finished dropping into siphon holes and now were on a one-way route to open water.

Ramshackle huts and cabins appeared downstream of the bridge, American flags raised, moldy old skiffs tied off. Plastic chairs. Propane bottles. Screened-in porches. Some had money and lawns, and others, backwoods shacks, looked ready to go with the next big storm.

Cormorants, enthroned in cypress trees like judges, peered down on us as the sky opened wide. The Aucilla spread into the Gulf of Mexico until it let go of itself, brown curls of water winding into milky brine. Fourteen thousand years ago, the river would have flowed for a couple hundred more miles across land now underwater. Sea levels have risen and most of the exposed continental shelf has been lost, the Aucilla truncated, half of Florida gone. Archaeologists have been mapping drowned river channels, sending divers who have found stone tools at sunken confluences and submerged riverbanks. Sites are mapped using side-scan sonar and multibeam bathymetry, which turns echoes into acoustic images. Shell middens or rock outcrops produce high backscatter arrays, which is where the drowned course of the Aucilla and its tributaries have turned up forty-five hundred pieces of chipped stone and broken spearpoints thirty feet below the surface, with more expected.

For people who knew these places when they were exposed, the sea must have seemed as if it were closing like a noose, the world being overrun, coming to an inevitable end if the water kept rising. Flood stories are deeply bound to Native American mythology. Between fifteen thousand and twelve thousand years ago, ice sheets rapidly collapsed and floods a hundred miles wide raged down the Mississippi, the outflow squeezed between Cuba and Florida, erasing most of the Bahamas as the water went for the Atlantic. This left the keys and islands we see today.

Sea levels rose in surges and did not go back down. The upsurges were not the inches and quarter inches we dread today, but feet and yardsticks, miles going under as people watched.

A breeze across this choppy water put brine on our lips as we glided over grasslands, over the backs of animals that once lived below this sea level. Under the water, mastodons returned to the river, wolves and big cats prowled the woods. I looked over my kayak, seeing people dressed lightly, slipping through the grass, atlatls raised as they went silently after a glyptodont, its round and armored back clearly visible from up here, its helmet-like head down and grazing.

We wanted to find an exposed shoreline to set up camp. But once the river fanned into Apalachee Bay, the land all but disappeared, turning into a tidal marshland, a low-energy coastal environment, hardly a coast at all. The forests we'd been paddling through were miles behind us, leaving only *hammocks*, islands of palms and cedars isolated across miles of salt marsh. With sunset coming on, we hugged the shore, or what passed for it, finding only doorways for alligators through stiff reeds. They were getting bigger, forming tunnels wider than my kayak. Some of the alligators might have been approaching Pleistocene size.

We spotted clumps of palms miles back into the sloughs, thinking we could land on one and find a place to sleep. The tide receded as we paddled inlets and outlets, a warren of islands and sloughs and still no land.

The sun peeled into an orange glow, lighting hammocks far away. The tide dropped faster, its current picking up against us, draining back into the Gulf. The chance of reaching a hammock escaped us as we floundered through a tidal maze, crusty banks of exposed oysters grating against our plastic hulls.

"This was a bad idea," said Pete, who was out of his canoe, standing in a shallow with his pants rolled up.

Alex stepped out in calf-deep water and said nothing, looking around, considering our options. Under a raspberry sky he pulled out the GPS, the first time the device had been used. The satellite map on the screen made it look as if we were folded into the convolutions

of a brain. We were eight miles northwest of the mouth of the Econfina River, and that was all we knew. GPS could not help us.

"Whatever we're going to do, we better do it soon," Pete said. That
was the ranger in him talking. He'd seen or heard of enough grisly,
stupid mistakes in wild environments. He could already imagine us
capsized, swamped in mud, trying to make it out of a marsh on foot
or by swimming in the company of large, unseen alligators.

"We could sleep in our boats," Alex offered.

Pete folded his arms. "I'm just saying we should make a call."

We'd failed to ask either Ed or Jim what to do when—if—we got
this far. I didn't want to sleep sitting up in a hard plastic coffin, not
with alligators bumping around in the dark. We picked a muddy
shoal and dragged our boats up, every step sucked out of the ground.
Leathery-sharp, black-needled rushes stood chest-tall, the ground a
sponge-like soil of organic remains, not quite land, not quite water.

Our two small tents went up. Alex and Pete kept their canoe close
and I put my kayak beside mine, tying it to a tent pole so that if the
water came up, the kayak and I wouldn't drift apart. We had dinner
barefoot, headlamps on, the sky indigo. Pete sat in a folding chair and
cooked on a little white-gas stove, blue circle of flame, battered old
windscreen wrapped around it, tools of the first people to land on this
piece of marsh in this century.

Mosquitoes were a nuisance, but not as bad as the swarms of biting, flying insects no bigger than poppy seeds. The worst of them
began to drop off as cool breezes came in off the water. With every
step and movement, we released a rich stink from the ground. Tiny
black crabs crawled across our feet as we walked to the edge of the
reeds, checking the tide, squishing down into the water. None of us
wanted a swampy night, to have to throw tents and gear into boats in
a midnight hurry. I thought of a prayer for an easy sleep, for sea levels
not to make a marked rise that night. But who was I to pray against
the sea?

Seven horses and one camel were attacked and butchered near a river
crossing in Alberta, Canada, 13,300 years ago, in a corridor of land
opening between ice sheets as Alaska and Yukon connected with the

Lower 48 in warming conditions. Butchered bones of a giant ground sloth, dating to between 13,738 and 13,435 years ago, were found in northern Ohio. The tip of a projectile made of mastodon bone was found embedded in a vertebra of a 13,800-year-old mastodon on the Olympic Peninsula of coastal Washington. A 13,400-year-old kill site called *El Fin del Mundo*—the End of the World—was excavated in the desert of northwest Mexico, where along with the remains of at least two butchered gomphotheres—long-jawed, four-tusked, elephant-like animals—spearpoints, flakes, scrapers, and an exquisite, completely intact transparent quartz fluted point were found.

The list of big animals being hunted or at least butchered keeps on, mammoths cut up in Wisconsin and Arizona, protein residues on points and butchering blades from across the country. A long-stemmed weapon known as a haskett, from the Western stem tradition and not Clovis, was recovered in Utah, with proboscidean proteins from mammoths or mastodons in its pores.

Matthew Hill, a zooarchaeologist at the University of Iowa, looked at sixty Paleolithic kills and carcass-processing sites from the middle of the continent to see what most people were hunting. Hill wrote, "In the low diversity grasslands of the High Plains and Rolling Hills, prehistoric groups hunted large game almost exclusively." Ninety-nine percent of the faunal remains he catalogued came from eight animals: bison, rabbits/hares, pronghorn, mammoth, turtle, bighorn sheep, deer, and prairie dog. The two largest, bison and mammoth, ranked near the top in numbers, partly due to their having the longest-lasting bones, and because they were either preyed upon or scavenged by humans. Hill concluded, "No matter what quantitative measure is used . . . large game dominates Paleoindian zooarchaeological resources." Of the Great Plains, that is. Smaller kills and smaller camps, which were probably the norm, have either not survived or rarely surface. The mountainous West and the West Coast appear to have been more diversified, stemmed and fluted points meeting among varied microclimates of taiga, grasslands, lakes, and alpine glaciers. In the Southern Rockies, eight kinds of projectiles were in primary use, from Clovis to stemmed points and varieties in between, and big animals were not their main focus. The Great Plains had a different hunting trajectory. Eventually, it would be known for bison hunts on horseback and woolly beasts shot from the windows of trains. During the Ice Age, the same land saw some of the biggest megafauna hunting forays.

Extinctions mounted at the end of the Ice Age. Most short-faced bear exited somewhere between 10,800 and 11,000 years ago, the same extinction range for many other North American Ice Age megafauna. A single *Arctodus* bone sample from Bonner Springs, Kansas, just west of Kansas City, has a date of 9,600 years old, evidence of the last of the last edging into the Holocene.

The extinctions happened at different rates, depending on the location. Wildlife ranges shrank as hunters pressed across the Great Plains in search of retreating herds. And environmental changes were everywhere. It wasn't just human expansion; a geologic era was falling apart. At the warming end of the Pleistocene, prehistoric Floridians watched entire landscapes disappear. The Gulf Stream stuttered in the flush of fresh, cold water melting from the ice caps, climatic systems becoming jumpy, the Ice Age coming to a wildly fluctuating halt over thousands of years. Rising seas gobbled up the mouth of the Aucilla as the faraway ice sheets retreated in a bluster of floods. Florida's mastodon savanna and pine mosaics went under so fast that dead trees would have stood like candlesticks half a mile across the bay.

Researchers examining the underwater courses of the Aucilla and other nearby rivers have found fossil mud beds. Their hard consistency indicates "especially prolonged desiccation," times of freshwater drought, sea level forcing up the water table, the last islands of savanna withering. Bone beds were also found underwater, broken teeth and fragments of the mandible and maxilla of a juvenile mastodon. Near the mastodon were the casts of dead trees, their roots upright, signs of their drowning still standing.

The sea rose and fell during the night without incident. High tide had massed under our tents, wetting the floors, but receded before filling in. I kept waking, reaching a hand out of my sleeping bag and palming the damp floor, making sure it wasn't becoming a waterbed. The following tide had already come in by sunrise. What had been muddy shoals and oyster reefs at sunset became, in early light, an almost drowned landscape. Chest-tall rushes rippled across the marsh like savanna grasses, so high we could barely see over them. Where we could find a view, the only thing out there was water and stranded islands of more marsh grass.

The nearest land was miles away, a lone hammock, an isolated rise of palms and cedars. It looked to be just a short paddle away. An hour later, pushing in with the tide, we were still trying to find a route, squeezing into passages where rushes rasped against the hulls of our boats through a maze of loops and sloughs. The palm-headed hammock couldn't have been more than an acre of land, close enough we could almost throw a rock and hit it, but we couldn't seem to get there, nor were there any rocks. With the narrower tip of my kayak, I led the way. The channel widened into a lagoon. At the edges of this deep-bellied hole, two alligators lounged on opposite shores. Both of them slid into the water, one armored snout lingering above the surface to watch me before it vanished. The shadows of their bodies glided out of sight, the water murky and briny back here. A Busch beer can floated in the rushes, carried in on the tide.

I called back, "A couple alligators in this hole, big ones."

"Alligators," I heard Alex say. "Understood."

I can't say I delighted in the thought of large swimming reptiles underneath me. But I wasn't terrified. On the Zambezi in Africa, where death by crocodile is common, I'd be scared . . . *Ice Age scared*. Here, I felt reasonably safe. These alligators wanted to get away from us, or so I told myself as I paddled above them. Below me, in the lagoon, they held their breath, not a bubble escaping.

On the other side of the lagoon, we stowed our boats where we could touch the edge of the hammock. This was an alligator island, their tracks lumbering across crusty mud. The mud had hardened into crackled saline, most of the palms dead and turned into bald pillars. Nine palms were alive, less than half of the original population. Cedars had died long ago.

This island is a sea-level marker where hammocks are becoming sterile decade by decade, life reduced to a crust, cedars practically mineralized where they stand, ready to become future underwater evidence of a changing world. As seas rise, saline waters push the freshwater table farther inland. Every storm moves saltwater up a notch. At an elevation of about two feet, our hammock had just a few more big hurricane seasons before seawater would take it for good.

Oyster shells had washed up on crusty ground where fiddler crabs darted from hole to hole. The ground was pearled with the muddy fecal balls they'd fashioned, as if turning this dry oasis into a balled-up sculpture gallery.

The surface was slightly moist and spongy, scraped with alligator claws, punctuated with the soft, rounded pads of a bobcat that came in from who knows where. Carefully, we moved into knee-high grass, watching for the snakes we imagined woven around us. Ospreys had been defecating in the barren cedar trees, branches and trunks scabbed with their white droppings.

I climbed one of the dead cedars, finding enough altitude to feel the sway of a stiff spring breeze flowing toward the Gulf. Rocking in clean air, this was the highest I'd been since we started along this nest of rivers. Until now, it was a matter of inches or a few feet above sea level. Fifteen feet changed everything. The larger land revealed itself. Inlets and tidal lagoons twisted around us. No wonder it took so long to find our way to this hammock: We were in a labyrinth, the far horizon a gray line of cedars many miles off, their columnar trunks stepping back from the Gulf as they have been doing since the end of the Ice Age. I felt as if I were clinging to a bowsprit, as some ancestor must have done, climbing up to watch horizons disappear.

You lost your world, I thought. The actual ground upon which you walked is now in the ocean.

8

CULT OF THE FLUTED POINT

13,500 YEARS AGO

Tink, tink, tink. The musical sound of stone on stone, marimba tones of an antler butt striking flint. *Tink, tink, tink.* Pieces shaped like angelfish flew to the ground. As the day warmed, the wall behind the flint knapper brightened with morning sun. Autumn chill eased off as he swept and swiped at a piece of Wyoming chert, a kind of rock found in a Clovis cache, carried at least three hundred miles over the Rocky Mountains, and dropped under a boulder near a creek in what is now Boulder, Colorado.

The knapper was in a valley in southeast Utah. Cliffs and eroded rock towers guarded the horizon. Cottonwood trees were golden. He'd been to Wyoming and Texas to fetch good rocks, pigeon-blood chert, and oily gray flint for making into weapons, a reasonable distance for rocks to be carried in the Clovis era. He studied the rock, feeling its heavy parts, its lightness. At his feet lay a mound of flakes from the hundreds of weapons and tools he'd made here. His hands were pressed and worn from a lifetime of banging on stone.

"You can't fool the rock," he said as he flipped it around in his hand. "You can a little, but not much."

Sitting next to him in a folding chair, I scribbled in my journal, recording the making of a Clovis point by modern hands. Greg Nunn is a rock sage, a philosopher of stone. A master flint knapper, he'd been doing this for most of his adult life. I'd commissioned him to produce a fluted Clovis point from the same chert found in the Boulder, Colorado, cache: tiger chert, a gray, mahogany color banded with dark minerals. He said he had his own source for it on a mountainside in Wyoming. He keeps it secret. Some of his toolstone sites are so precious he's filed mining claims to protect them.

Nunn sat on a portable cottonwood stump in front of the south wall of his house, a spot that he preferred. He never liked working in winter and thinks that Ice Age people would have agreed. The rock doesn't respond well to cold, he said, and it's a nightmare on your fingers. Autumn is good, a time of hunting and warm, ethereal days, sunlight becoming long and rich.

Already he was bleeding a little. You have to get used to it, he said. Every good point, you're going to bleed, especially Clovis.

"These are aggressive strikes," he told me. "You just have to hog it out, make it or break it."

The Clovis point, which seems to have developed first in the East between Florida and Maryland, was part of an almost codified toolkit of bone rods, preforms, scrapers, blades, and nearly identical flint-knapping tools. The utility of the bone rod, which is usually beveled at both ends, remains a mystery; it could be a weapon, or the handle of a butchering tool, or the shaft for a projectile. This collection of artifacts may be the equivalent of the tools and weapons traditionally used by Indigenous whale hunters in the Far North. In Point Hope, Alaska, early Inupiat whale hunts use the *qalugiat*, a lance made of a sharp piece of jade mounted on a long bone that is attached to an eight-foot wooden handle. This is used to kill the whale after it has been captured, by driving the lance in close to the heart, or liver, or lungs. Once the whale is dead, a cutter called a *kaugaq* is used, made of slate or jade, the blade a foot and a half long and three feet wide. This is for butchering. The hunters carry a pair of knives made of bone, one with a long, thin blade, the other with a rounded blade like a spatula. The long blade scrapes the sides of the boat where ice collects and the other is for digging underneath the accumulated ice on the boat.

The ancient whale hunt is a ritual, preceded by prayers and supplications to the new moon. Tools are readied and sharpened, and the hunters stay in a different place than their families. The Clovis mammoth hunt may have been similar, the weapons and tools prepared in ceremonial fashion, rubbed with red ochre until the beautiful stone could not be seen, bathed, as it were, in mineral blood. People quarried ochre back then, making cultural sites where they gathered the mineral from the ground. Clovis points without visible red on the surface have microscopic ochre traces in their cracks and pores. An archaeologist tells a story about taking a Clovis point to classrooms where hundreds of children handled it. No matter how many kids touched it, he still couldn't wear a white shirt without getting red on it.

Nunn hadn't made a Clovis point in a few years. He came out of retirement for this one. Knocking hard and snapping off pieces, alternating between using a rock cobble as a hammer and then the back end of an antler, he said, "I'm replicating a process that's part of a technology." He hopped the dark rock in his hands, saying that he saw Clovis less as a people than as an industry, a way of shaping weapons. People speaking different languages may have been engaged with these weapons in a way that transcended North America's many geographies.

"I'm not hunting mammoth, but I'm mimicking what I see in the rock," he said. "I'm not a Clovis person, I am understanding Clovis strategy. I can see they were deliberate, methodical. They were thinking way ahead. They knew it was going to break, so they designed them so they knew *where* they were going to break."

For several minutes at a time, Nunn seemed to fly through the rock, taking it from a big tab in his lap down to a teardrop shape the size of a serving dish on his knee. He slowed, struggling with a crystal core, a dense lens the size of a penny inside the rock. He blamed himself, saying he'd seen this coming but hadn't worked around it correctly.

"Now, it's a problem," he said.

Turning the rock and banging from one side, turning and banging from the other, he told me that if he hit it too hard or in the wrong place, he'd take off too much and have an even bigger problem. Too gentle, and the problem would become even more intractable, options running out. You find surrenders like this stone all over the country, half-blanked and ready to be made into something else, but discarded; maybe something better came along, or it was roughly fashioned at the moment, out of necessity, and then dropped and left behind. This is how people seeded the land, working through stone for thousands of years, soils glittering with the flakes they left behind.

Nunn is not a scientist. He is a hunter. He makes his own projectiles for taking down deer and elk. His experience on the land surpasses that of most archaeologists, who make weekend field trips and spend most of their professional lives in data and papers. Nunn knows the country by its shapes, how it breathes animals, where it shows the way and where it does not. He grew up on a ranch in this part of Utah, where his father was also a knapper. Nunn inherited the curiosity, finding broken stone artifacts in the desert and replicating them. He made a profession out of knapping, learning every stroke and turn of whatever technology he handled, classic American points, Danish daggers, fluted Clovis projectiles. Archaeologists come to him to understand the process, learning about knapping techniques and how rocks were moved, where they came from, where they went. For research, he visited collectors, rock hounds who picked up spearpoints and arrowheads in the field, and he said, *Show me whatever you've got.* He'd run his fingers through flake scars, turning each artifact, seeing the ancient work with hammerstones and burins, the angles of their blows, the force of the strike. Clovis he thought of as a code, not an easy one to break. "It was meant to be hard," he said. "Not everyone could do it."

Clovis was a way of working stone in an almost identical fashion from coast to coast, the same general morphometrics, like Pleistocene golden means found in Maine, California, Oregon, Florida, and places between. This implies connection, communication, a continent held together by a single technology 13,500 years ago.

A fluted Clovis point employs a technique called the "overshot flake," a blow precise and hard enough to knock out a long, slivered wedge from edge to edge on the rock. Done imprecisely, the rock breaks in two, left useless but for making into blades and scrapers. Considering how far Nunn has to go to get his materials, and how guarded his tiger chert source is, he dislikes wasting rock. He said, "You see plenty of Clovis points without the overshot flake." This, he admitted, was probably an excuse for old, tired, or inexperienced knappers who didn't want to risk their rock.

Nunn lined up his piece, etched a platform into its edge, swung a couple practice strokes from the pitted knob of moose antler, and struck with a hard downward arc. An overshot flake flew from the backside. He practically caught it with his hand. Flipping the thinned teardrop around, he studied the shallow trough he'd blown across it.

"That was a good one," he said.

Settling his body back to work, he said, "I don't want to get too mystical about this, but a rock has power, or spirit, whatever you want to call it."

A 1988 ethnographic study of hunter-gatherers in far northern Australia found that certain kinds of rock have cultural significance. People walked hundreds of miles to get them. Lester Hiatt, the late scholar of Australian Aboriginal societies, wrote, "At the quarry, the men spoke of the stone growing up in the ground. Only here at Ngil-ipitji did true 'killing stone' grow. The cross-sections of weathered rinds were compared to that of a kidney, with the best interior stone of pinky-grey silcrete referred to as *djukurr* or 'fat.' An esoteric oblique meaning of this word is power. It is this mystical power derived from supernatural sources integral to the site that gives the Ngilipitji stone blades their stupendous killing force. Once struck, man or beast is doomed."

Hiatt wrote that the Ngilipitji stone blades were "prized and feared" and that they formed a key resource in a system of trade that helped integrate distant tribes into "a single cultural complex."

This is what Nunn and most archaeologists see in the Clovis age, a single cultural complex across the Americas with stupendous killing force, the rock for their points sometimes gathered from hundreds of miles away, often chosen for color and pattern, some of the hardest rock to work selected because it was visually striking. If you are going to kill, do it beautifully, powerfully. Choose the right stone.

In 2009, landscapers installing a small koi pond at the residence of Pat Mahaffy in a hillside neighborhood of Boulder, Colorado, found a cache of Clovis artifacts — eighty-three stone knives, blades, blanks, and flakes that would date no later than 13,500 years ago. The rocks had been carried to the site from geologic sources up to three hundred miles away. Based on what is called "bagwear" on the rocks, it appears to have been done in a single journey.

The journey was from an upper corner of Utah, through Wyoming, into Colorado, and across the Rockies to the Front Range, with key resources picked up along the way. A sparkly butternut-colored quartzite was collected around the Uinta Mountains in northeast Utah, probably the first stop on the route, followed by an abundance of flakes and worked chunks of tiger chert, also known as Bridger

Path of rocks carried from their geologic sources through
Utah, Wyoming, and Colorado to the Mahaffy Cache

Basin chert, from the Wyoming-Colorado border just north of Dinosaur National Monument. This tiger chert is the same handsome,
striated, nut-brown rock that Greg Nunn used to make his Utah point
for me—hard to work with, but appreciated for its banding and rich
colors. Next is a milky, blue-white Kremmling chert, a fine quartz-
like rock from the high elevations of Middle Park in north-central
Colorado, just west of the Continental Divide from Boulder.

The movement of rock defines the sphere of human reach, in this
case following rivers, and passes through Utah, Wyoming, and Colorado. The Clovis age moved jaspers, cherts, and flint. The places they
picked them up, then stopped to work them and ultimately put them
down, reveal their paths.

Doug Bamforth from the University of Colorado arrived at the
Mahaffy home after landscapers had pulled out about eighteen
pounds of imported rock from underneath the boulder where it had
been hidden since the late Pleistocene. He found evidence of the
same transport wear on both sides of the artifacts, meaning they traveled together, rubbing against each other in a relatively swift journey.
"This suggests that they kept up a steady pace," Bamforth said. "They
didn't stop for a couple of months along the way."

The cache in Boulder shows a west-to-east movement of rock, but the bigger trend found in at least twenty-one North American caches shows rocks moving southeast to northwest. Edwards Plateau chert from Buttermilk Creek in Texas ended up in Oklahoma and Colorado. An artifact of white chert that originated near St. Louis, Missouri, turned up in a central Washington cache two thousand miles away. Technology seems to have moved with older Clovis finds in the East, and more recent artifacts in the West. This curve travels from Florida up to Alaska, where fluted points twelve thousand years old have been found near the diminished shores and islands of the Bering land bridge in western Alaska. This appears to be a splash-back, a movement returning to where people came from. Clovis influence went south, too, with fluted points found from eleven thousand years ago near the tip of South America, including a curious stemmed-fluted combination found at Cueva Fell in far southern Chile.

The lay of most Clovis caches on the continent forms a pattern, a swath from Washington to Texas, more finds on the eastern side of the Rockies than the western, none documented on either coast, and none in the Southeast, although Clovis existed in all of these places. Most of the caches are near the inner-continental heads of long rivers, as if the entirety of North America was their playing field, the Rocky Mountain chain a convenient locator near the middle.

Some archaeologists maintain that these distances were a matter of trade and not single journeys. Having walked the inner parks of Colorado, watching mountains appear ahead, pass by, and disappear behind me, I believe they could have it wrong. The land is easy to cross.

Now a passenger train crosses the Rockies along the Mahaffy route. The *California Zephyr* is one of the only contiguous railroad passages over the mountains, a spectacular trip from the observation car. The route takes advantage of an ease in the terrain, broad, open parks between mountains, and a river with headwaters to lead the way up and over. The train incidentally follows Clovis people who came through the last chain of mountains, where thirteen thousand years ago a passage had opened between retreating glaciers, connecting the west and east sides of the continent.

Though the gorges are beautiful, the upper Colorado River wrestling far below the train, the most epic scenery on the *Zephyr* comes when the journey tops out. After a hundred miles of mountainous terrain, the Rockies end in a four-thousand-foot plunge to the Great Plains, the flat American interior. It feels like being hurled into the sky, fired out of the cannon of the mountains. Cities and towns spread across the plains, and rivers wind eastward, turning eventually into the Mississippi, leading to the Gulf of Mexico more than a thousand miles away. This train is my regular commute when I need to get across the state from my home on Colorado's Western Slope to, say, a major airport on the state's Front Range. I am following a groove laid down thousands of years before me, as if every footstep falls inside another. Even if we can't see it, we can bet we've been here before.

The Clovis party carrying these rocks would have topped out at the crest of the Rockies above Boulder in the same place. Bamforth thinks that this would not have been a residential group, but a foray sent out to procure the raw materials and return in one shot. The area around Boulder does not have good toolstone. Fetching it may have involved a six-hundred-mile round trip to the corner market, bringing back reduced pieces of Wyoming tiger chert, Utah quartzite, and a pale chert from the upper Colorado River.

"I'm going to Boulder, and I know there's no decent rock out there," Bamforth said. "I might just make sure that we're not stranded."

"Stiff little rascal," Nunn said. The tiger chert was giving him a hard time. I had asked him to make the point with this specific rock from Wyoming, a type carried by the Clovis party over the mountains to Boulder, and buried in what would be Mr. Mahaffy's yard. Hard but pretty, he called it. Not his favorite or most responsive rock, but pleasing to the eye.

He'd selected this piece from his rock stash on the first morning, saying, "This one's got potential." He sat down on his cottonwood stump and started scratching and pounding. "Let's see what it looks like inside," he said. Big pieces flew, a third of the rock shed in just a few blows. "That got rid of some mass," he said.

Flakes came off cleanly, like he planned. He reached down and picked up the best ones, fitting them back into place to admire the accuracy with which they separated.

He said a fluted Clovis point is a high-risk weapon. Personally, he thought it wasn't worth the effort, unless all you wanted to do was make a specialized instrument, a beautiful mammoth killer, something that would stand out in the world of spearpoints and atlatl tips.

He banged down with a cobble and didn't get the flake he wanted, only a puff of impact dust, smell of ozone.

"You can hear the difference, your ears are as much a tool as anything," he said.

He took another swing.

"You hear that, it was dull, not as sharp."

On the third strike, the stone rang like crystal. A flake popped off the back. He picked it up from the ground and fit it back in place. "I could have drawn that with a pencil," he said.

By the end of the second day, he had this ten-pound block down to a ten-ounce shape, a flattened missile. He apologized for how long it was taking. He said he would have done this in one day in his prime. A Clovis person might have finished the work from big block down to sharpened point in a few hours.

As the spearpoint took shape, he used smaller and smaller tools, down to the point of a mule deer antler and a stone burin no bigger than a dice for scratching out the edges, making tiny platforms that he knocked off with a rounded river cobble, chosen so it fit like an egg in his hand. "These are the real treasures," he said. "A good hammerstone is hard to come by."

When the afternoon sun dropped through the cottonwoods, his wall no longer in the light, he laid the almost-finished point onto the flake pile and we went inside for a beer. He told me we'd get back to the projectile in the morning.

"There's no doubt in my mind they had big gatherings, big hunts," Nunn said at his kitchen table. "They were probably doing it long before Clovis, but Clovis is just when you start to see it."

He said if he were hunting, he'd never use a Clovis point; it's too much work, they're worth more sold to collectors than broken in the field. They are bought in cult-like fashion, this old design still all the rage, but still it was too much work, he thought, for the hunt. "I'd use a simple biface," he said. "Not one of these."

A cache at the Crowfield site in southwestern Ontario contained 182 stone artifacts, some of them still functional and others intentionally burned and destroyed. This has been taken as evidence of a "sacred ritual." These were likely seen as more than just weapons.

Magic may have been imbued, special killing stones gathered for their power and fashioned in strict tradition.

The final touch is the "flute," a channel knocked up both sides, starting at the base like a stamp, a chop. To be catalogued as a true Clovis point heading into action, not a preform waiting to go, it needs the flute.

On the third morning in warm, clear sunlight, he popped fluted channels into either side. They were meager flutes, just enough to mark it as a verifiable Clovis hunting weapon.

Nunn wasn't impressed with his final product. Tiger chert isn't easy, he said again. To me, it was gorgeous. The projectile was thin as a letter envelope, and strong. It sounded like a musical instrument, the bar of a xylophone, as he placed it in my hand.

"It's a Clovis point," Nunn said. "I've seen worse."

9

THE LAST MAMMOTH HUNT

I sat with a friend on a dune crest, nothing else alive nearby but the sharp flares of a yucca sticking through sand. In an early December chill, we dug our bare feet into bone-white dunes. This desert forms a great blankness across southern New Mexico, visible from space. A lake used to be here, two thousand square miles of crystalline surface against a shoreline of wooded mountains on one side and pure white gypsum beaches and grass on the other. Driven by heavy rainfall and meltwater catastrophes out of the Rocky Mountains, glaciers collapsing at a pace not seen for more than ten thousand years, the Rio Grande regularly jumped its banks and spilled through intermountain basins, filling this lake. People left artifacts around its shores, not many, but enough to tell of the presence of fluted weaponry. Stemmed points are plentiful along intermountain lakes hundreds of miles north of here, but this looks more like Clovis territory, an Ice Age outpost. Dunes once rolled into blue water, like the sand dune lagoons on the north coast of Brazil. The place is now pocked and sun-baked. Mountains on the other side of the basin roll onto their sides like capsized battleships, the forests that once grew there gone. A missile range lay between here and there. Several miles to the south, a black plume of smoke rose as if from a naval battle in a dry, alabaster sea. A bomb had detonated and something had caught fire, a dark fist of smoke thrust into the calm blue sky.

Clovis people were used to environmental catastrophes, but this—

the land turned alien and struck by human-made bombs—was not something they would have easily grasped. The end of any ice age is violent and unstable, giant floods, epic catastrophes. But how do you explain a country laid bare? The same shapes of mountains were here, the same contours of shorelines, but everything else had changed, down to airplanes blinking through stars and the nearby town of Alamogordo, New Mexico, lighting a corner of the sky at night like hellfire.

Nick sat beside me watching the smoke plume rise a couple hundred feet. He'd been a bomb tech in the war, Iraqi deployment, and worked at the White House for a thirty-day stint, alone in a park wearing the "Hurt Locker" suit as he inched around an abandoned briefcase while a news helicopter circled overhead. He told me it was one of the more riveting moments of his life, just him, a potential bomb, and all the world watching. We became friends when I taught at the University of Alaska, where he was a writing student. He came down to New Mexico to join me on a desert mission. I was looking for mammoths. He focused a pair of small field glasses on the explosion, its source hidden behind sand dunes. We'd been hearing atmospheric concussions all day, faraway explosions, surface-to-surface missiles launched from one end of the range to the other. He said the smoke could be anything, a personnel carrier, a tank, the shell of a helicopter suspended aboveground on a cable as a target. He described to me a super-dense warhead made of depleted uranium, not an explosive but a simple spearpoint for ripping through an armored tank, going in one side and out the other in a split second. "It pierces a tank and leaves a hole the size of a baseball," he said. With his hands he showed me how big the hole was. Anything not solid, like soldiers, would be instantly fired out of the tank through this aperture. He'd seen it done with a sheep. The animal was liquefied, and sprayed across the ground.

"Then you can get in and drive away, no damage to the tank," he said.

When we toured a military museum at one of the White Sands bases, wandering through a sculpture gallery of retired weaponry— missile launchers, rockets, and the white, bolted orbs of early atomic bomb casings—he started to laugh, saying, "This is what we do better than anything."

"Killing?" I asked.

"Killing," he said. "Hell, yeah."

He was thinking M16 rifles and hair-trigger bombs made to look like children's toys. I was thinking Clovis. Different parts of the same story. Thin, strong projectiles made of stone thrown silently through the air increase the sphere of the hunter. Small and scrawny as we are as a species, we have the influence of giants.

Nick and I were accompanied by an itinerant mountaineering guide named Charlotte, and my buddy Jordan, a tall, lanky photographer from my Colorado hometown. Our sphere of influence felt like only our footprints, which blew away quickly in the wind. I wasn't the trip leader. I explained this up front: This was an exploratory mission, an arcane form of nomadism, no hierarchy, just pick up and go. We would move by consensus along the axis and edges of dunes, tents dropped between gentle swales and crests. I told them that if we ran into a fence with a sign reading "U.S. Government Property No Trespassing," which would mean the edge of the missile range, I would not go over it. They could do whatever they wanted.

Nick handed the binoculars to me, and I returned to scanning the blasted landscape beyond the dunes, the blistered lakebed up against the mountains. Mammoth tracks had been discovered out there, off-limits. I'd tried for two years to get permission, but the chain of military command denied the permit somewhere higher up. Instead, I walked the edges of the range, finding old bullets shot from airplanes, ground sprayed with old practice fire, sheer chance that tracks would ever survive out there.

The tracks were made by Columbian mammoths walking along wet, shallow shorelines, according to the paleontologists who worked on the missile range. In photographs, they look like circular ripples on still water, dabs of memory placed left, right, left as the mammoths ambulated. The prints are not stone. They are made of soft sediments, slightly more resistant than the surrounding plain, the last part to erode. In a sense, they are still fresh, filled in and buried, but still soft. Once exposed, they don't last long, sometimes years, sometimes months. One researcher returned to record a trackway he'd spotted a year earlier, and it had worn back into the gentle pan of the missile range, barely visible. After waiting tens of thousands of years, they come to the surface and are gone as swift as a spark.

Some of the tracks that have been found are from isolated individuals, and some are moving together as families, or perhaps the mammoths followed the same route at different times, much as African elephants do today with their long, deeply trodden trails to water. Some tracks are subtle, the shapes of big pancakes domed slightly from the ground. Others rise proudly, sprouting like mushrooms from the crater-riddled playa. Impressions of toppled trees have also been found, pushed over and denuded with branches broken off, like those left behind by modern African elephants, only these animals were a few tons heavier, pushing over larger trees to feed on the upper branches, peeling off large sections of bark in the lean seasons.

The hind paw print of a short-faced bear was documented here, toes and claws pushed off as it moved over plaster-like mud. Across what looks like an unending wasteland are hundreds of smaller, criss-crossing stabs, marks of animals that weighed less than a ton: dire wolf, horse, panther.

When I looked beyond the dunes into beige oblivion I pictured these animals moving along the shore, grazing and hunting around what were once grassy beach dunes. Their winter chill would have been sharper, but not unbearable, like the windy rawness of New York in winter. We'd gone barefoot in the morning, temperatures just under freezing, to let our soft arches and the nibs of our toes acclimate. Paleolithic people's feet would have been hardier. Even if they wore sandals and leggings—perishable items that have not survived—they would have been in constant and direct contact with their world. No doors to close behind them. Every night was stars.

Sliding my feet under the sun-warmed upper inch of sand, I put the Ice Age back in place, seeing its hulking animal shapes in the distance. I thought of Nick and I as two hunters pausing atop a white beach dune. The mammoths were a few miles away, their humped shoulders and heads moving along the lake, tusks rocking back and forth to the rhythm of their stride.

Fourteen mammoth and mastodon kill sites have been documented in North America, most of them associated with Clovis toolkits and projectiles. The number may seem low, but it represents only those where human agency is irrefutable—not just scavenged but attacked with weapons. Countless more must remain undocumented. Europe, with seven hundred thousand years of closely studied early human and hominid habitation, has turned up twenty-one proboscidean kill sites, and Africa, with more than a million years of big game hunting, has only eight prehistoric elephant kills. The fourteen found in North America stand out as a high number for a much shorter occupation. In their 2008 article, "How Many Elephant Kills Are 14?," archaeologists Nicole Waguespack and Todd Surovell called this "the highest frequency of subsistence exploitation of *Proboscidea* anywhere in the prehistoric world."

Waguespack and Surovell sampled faunal remains from thirty-three Clovis sites across the country to see what the Clovis might have been hunting and butchering, and discovered that on average the largest species were found more often, e.g., more mammoths than hares. Rather than exploiting all the prey they came upon, these Clovis hunters were going for the largest animals. The thirty-three sample sites are almost exclusively top-heavy, mostly megafauna with little else. The two archaeologists have been accused of cherry-picking their sites, but the abundance of large bones still stands out. Perhaps not everyone was hunting and eating proboscideans, but in some places or at some times, that is exactly what they were doing. Smaller bones of snakes or deer would not have lasted so long, skewing the results, but these accumulations of hunted and butchered megafauna bones represent unique events. The two researchers conclude that people at these sites were engaged in "the extensive and selective use of large-bodied prey."

"This does not mean that Clovis existed by mammoth alone," they wrote, careful not to stir the hornets' nest where some say Clovis were dedicated mammoth hunters and some say they certainly were not. But with their data, they had to conclude, "based on estimated encounter rates, Clovis hunter-gatherers often ignored opportunities to harvest smaller game species, likely in favor of obtaining a higher-ranked resource."

Bruce Bradley, from the University of Exeter, considers Clovis and its penchant for hunting the largest animals a kind of "cult." He points out that fluted technology spread from coast to coast within two hundred years. "That's too fast," he told me. In the Paleolithic, at a time when tools could remain unchanged for tens of thousands of years, this would have been breakneck speed. Compared to historic models for birth rates and migrations among hunter-gatherers, the

Clovis age looks like an explosion. Similarly fluted points made it as far as the bottom of South America and to the edge of the Bering land bridge in Alaska, with one possible specimen found in Siberia, making the land bridge a two-way street. Something more than hunting efficiency carried these fluted points to the ends of the New World.

"How do you become a man if you can't find the mammoth?" Bradley said. "I think the whole Clovis thing out in the Far West was entirely going after mammoths, looking to get that status. We don't find that many Clovis settlements in the West. They were coming out here looking for that last population of mammoths so they could fulfill their religious duty. That's where they're encountering the local people and passing on their technology."

Bradley, who has made a career out of the spread of lithic traditions, especially the unprecedented rise of Clovis, believes this new big-game projectile is a sign of a much larger ideology that took hold faster than technology should have spread, thus his word "cult." The first wave of human arrival—barring Steve Holen's evidence of early bone-smashers and Dennis Stanford's Solutrean incursion—is by West Coast around sixteen thousand years ago. The second was a later influx around Clovis time when an ice-free corridor linked the Far North to the rest of the New World. Growing in population, people mixed, connecting with each other, reaching a continental zenith at a pivotal moment in climatic history.

While sixteen thousand to fourteen thousand years ago had been ideal for human arrival—plenty of water, the cold slipping away, permafrost leaving the ground, grasslands rich in megafauna—Clovis arose when the good became too much. Ice dams were bursting, shorelines disappearing. Thirteen thousand years ago, a glacier dam broke in Montana and took out half of Washington State and some of Oregon and Idaho. Ebullient masses of mud and glacial silt carried off bloated mammoth carcasses and archipelagos of drowned horses, everything inundated but the highest buttes and mountains. One waterfall was three and a half miles wide, water plunging four hundred feet into a cavity carved suddenly from the earth. Large portions of landscape were erased, giant gravel bars laid across the Pacific Northwest, sculpted into ripples twenty feet tall. Both Clovis and stemmed artifacts have been found in the area, meaning that this flood would have been witnessed by human beings. If anyone survived it, they must have sought shelter, perhaps atop a basalt-capped butte in the middle of the flood, the rock shuddering underfoot as everything turned into crashing debris as far as the eye could see.

It wasn't just floods. In Central Mexico, volcanoes were erupting, encasing mammoths and human artifacts in superheated igneous ash flows, giant corpses cooked and tumbled all over each other. A controversial comet impact may be in the mix, too, with glass spherules sprayed around the Northern Hemisphere, molten silica found spattered against plant and animal remains. One hypothesis is that this atmospheric impact, which struck over North America but spread worldwide, changed the planet's climate for the next thousand years. Images in a carved stone pillar at Gobekli Tepe in Turkey show a catastrophic and long-remembered event, a swarm of fragments falling from the sky, possibly changing the planet's rotational axis. Star patterns recorded at the site pinpoint the event to around thirteen thousand years ago.

As if Earth weren't taking enough of a cosmic smack, some stellar event left a blast of high-energy photons and possibly lethal UV radiation in marine cores and tree rings about 12,830 years ago — either a core-collapse supernova nearby in the galaxy, an immense solar flare from our own sun, or remnants of a bolide impact to the atmosphere. These events, together with thousands of years of epic flooding, dramatic sea-level rise, and the swift decline of charismatic megafauna, would have caught people's attention.

Bradley believes that the appeal of Clovis was largely in reaction to all of this. People may have created what he calls an "apocalypse culture," a way of standing tall in the face of enormous changes. He and his colleague Michael Collins, who heads research at the Gault site in Texas, wrote, "When the societal stress became too much and there was cultural distortion, either a new system would be needed or the society would fail and disperse or become extinct." Clovis was the new system. Bradley and Collins see a new American order in flaked stone technology, perhaps the shadow of a larger set of invisible behaviors and symbols, a codified system they describe as being "'designed' to bring order, purpose, and meaning to life in the new environment."

Bradley jokingly called this "Mammoth-Feast Destiny." Not entirely a joke, he used the term to describe a cultural transformation, a mode of proliferation — as if people were taking down megafauna as a way

of making themselves feel larger, better able to face these huge, cha-
otic events around them. Some scientists take his hypothesis seriously
when considering the impact of environment on a culture, while others say it's "drinking the Kool-Aid," and that Clovis was just another
technological complex. But it wasn't. This was something new and
grandstanding, and it spread swiftly from sea to shining sea.

Lava flows, dry arroyos, and the edges of sand dunes course through
the basin of the Jornada del Muerto in southern New Mexico. White
Sands lies just to the south, in the bottom of Tularosa Basin, home
of Ancestral Lake Otero. The Jornada lies to the north, a big country, one circular horizon rumpled and broken with mountains nearly
barren.

In English, the name Jornada del Muerto means a day's journey for
a dead man. In the summer of 1670, a Spanish party passing through
the broad, arid expanse found the remains of a dead horse tied to a
tree. Nearby were the bones, hair, and clothing of a man. The name
stuck.

Mountainous escarpments stand apart from each other in the distance, gaps opened between them, gun sights on the horizons. Near
one of these gaps, which looks toward a once-glimmering and now
empty lake, archaeologists have unearthed the largest, most deeply
stratified Clovis site in the Western United States. Known as Mockingbird Gap, this was a tool-prep camp created when the Jornada del
Muerto was a grassland, the head of every arroyo moist and trickling
with springs, mammoths wandering the nearby lake edge.

Mockingbird Gap is one of the gun sights on the southern end of
the Jornada. The thirty-five-acre site was occupied broadly and repeatedly, producing a rich assemblage of Clovis points and unfinished
points, knives, gravers, endscrapers, sidescrapers, flakes, cores, and a
single mineral bead drilled through the middle, probably a piece of
body ornamentation. This was a manufacturing station for the hunt,
and over time personal effects would have been lost, a hunter checking leather pouches, wondering where that red bead got off to. Bruce
Huckell, an archaeologist from the University of New Mexico, co-led
an early 2000s survey and excavation of Mockingbird Gap with Vance
Holliday from the University of Arizona. In a 2006 investigation at

Mockingbird Gap, at least a dozen individual sites were located that "represent either discrete campsites of small social groups of Clovis that came together for short periods of time or perhaps repeated occupations of the site by a single group." This was an early American aggregation, unusual for Clovis people, who tended to use fire pits only once, a quick layer of ash and charcoal and they were gone. The land was gradually becoming known, people returning deliberately to places they'd been before.

I put the question of population and movement to Doug Bamforth at the University of Colorado, who said, "One of the big stories of Native North America is a really tiny number of people ending up as a lot of people, and that takes time." Bamforth sees them covering large distances on foot, which is how Clovis spread so fast. The placement of other groups and of resources defined how they filled the spaces, setting routes across horizons to get the contact or materials they needed. Bamforth said, "For the first few thousand years, hardly anybody was here. How do you live? How do you connect with other people? You get up and move."

Paleolithic Americans carried rocks hundreds or thousands of miles, distances most of us only consider from planes or cars. Their mental maps were on a continental scale. With the appearance of caches, people had to have had some knowledge of where they were going and when they might be back. The continent could have been a single playing field, waves of mountains and plains from coast to coast traveled by a people spreading their fluted points as far as they could reach.

Studying the JS Clovis Cache from the Oklahoma Panhandle, University of Oklahoma archaeologist Leland Bement wrote that the weapons in his survey had been "used hard and put up dirty." Like JS, many caches appear to have been used again and again, points broken and refashioned, signs of action in the field. Other finds consist of untarnished points lined up and ready to go, many not yet knocked with the diagnostic flute, brand-new and still in the box. The country was being seeded, hunting supplies made ready, some of them used, some not.

Clovis was far from alone; stemmed-point users dominated the West Coast and the pluvial, mountain-and-basin interior of the West. The different points mingled, but mostly stayed apart, as if representing two distinct continental cultures. People with different technologies clearly knew of each other. Stemmed points have turned up on the Gulf Coast of Texas deep in Clovis territory, while a fluted Clovis point was found during the construction of a celebrity's home in the capes of Malibu, California, a stemmed-point stronghold. The two met in the middle, one arriving before the other. A pair of Great Basin archaeologists, Charlotte Beck and Charles Jones, have written that "initial colonization of the intermountain region most likely involved groups moving inland from the Pacific carrying non-Clovis technology, which was already in place by the time Clovis technology arrived."

Eight miles from Mockingbird Gap in the dry plain of the Jornada del Muerto is the Trinity Site, another peak in the history of American weaponry. A sandy blast-zone lies 340 feet across. At the center is an obelisk of black and mortared rock with a plaque that reads:

> **WHERE**
> **THE WORLD'S FIRST**
> **NUCLEAR DEVICE**
> **WAS EXPLODED ON**
> **JULY 16, 1945**

The crater is still above radioactive baselines for healthy, prolonged human exposure, but plumes of sacaton grass and snakeweed have grown back, the Jornada covering its wound.

The bomb used at the Trinity Site was a wire-covered sphere just big enough to crawl into if it weren't nearly solid, with a plutonium core set within an explosive shell. Its active components were built in a government lab at Los Alamos in the Jemez Mountains of northwest New Mexico during the final campaigns of World War II. Then they were brought down here for assembly and detonation.

The explosion in the Jornada happened before dawn when the stars were still out, clouds passing after a brief rain. The bomb is said to have been as bright as day, mountains miles away completely lit by it.

If the first people had seen the explosion from Mockingbird Gap, they would have shielded their eyes. What would they have thought, seeing this second sun rising? By Clovis age, with all they'd gone through, giant floods and regional extinction, continental shelves swallowed by the sea, they must have followed this mushroom cloud upward, necks craning, and wondered, *Now what?*

I remember Bruce Bradley describing the biggest Clovis point, two feet long, saying, "You can't use that for anything." The same might be said of nuclear weapons, too big to be truly functional, more a threat and a statement than a handy projectile.

Half of the toolstones at the Mockingbird Gap Clovis site come from local green-black cherts and a reddish jasper. Other toolstone was brought down from northwest New Mexico, two hundred and fifty miles away, sourced in the Chuska Mountains on the Arizona border, and New Mexico's Zuni Mountains. This gives some idea of how far these people would travel. One piece of obsidian found at Mockingbird Gap was traced to the Jemez Mountains near Los Alamos.

People gathered and processed stone in northwest New Mexico, brought their blades and preforms here to the Jornada, and worked on them further at camps that may have been launch sites for megafauna hunts in the marshes, arroyos, and lake edges.

Pieces of the bomb used at the Trinity Site were prepared in the same part of northwest New Mexico and were brought to a ranch on the Jornada, assembled there, and detonated at a conveniently remote location. If only an anecdote, the pattern of resource collection, assembly, and deployment are the same at the Trinity and Mockingbird Gap sites thirteen thousand years apart. Picture scientists studying the Jornada in the distant future believing that there'd never been a hiatus between the Ice Age and now. These bombing ranges would be an unbroken carpet of human history, an evolution of weaponry where once we killed mammoths, then we killed everything.

The scientific name "Clovis" comes from the town of Clovis, New Mexico, near Blackwater Draw in the eastern part of the state. It was the first Ice Age kill site in North America to be identified, what some scholars see as the first evidence of a distinctive people on the continent, with a codified way of making projectiles carried with toolkits of bone rods, knapping instruments, and stone preforms ready to fashion into whatever the day required. The fact that the name came from New Mexico is happenstance. It could have come from any town in any state, Clovis having been found just about everywhere.

In the late 1920s, the Dust Bowl scoured enough ground from the High Plains that the desert around Blackwater Draw revealed the bones of megafauna, Columbian mammoth femurs like tree trunks mixed with the remains of giant bison and charismatically fluted weapons. At the time, this would have been the first scientific proof that early people in North America coincided with Ice Age megafauna, a fact previously disputed. The going theory was that humans appeared late, landing fully formed as Native Americans.

Scientists were alerted to the presence of these bones and weapons in 1929 by a nineteen-year-old Eagle Scout named James Whiteman, who went by the nickname Ridgely. From the town of Clovis where he lived, Ridgely began sending letters to museums and scientists, telling them he had found "extinct elephant bones," which turned out to be mammoth bones, along with horse, camel, bison, and other extinct species. The young man also reported "Indian warheads," his interpretation of fluted projectiles. He had some idea of what he was seeing in Blackwater Draw, having visited the same museum in Denver that I did several decades later. He'd also gone to the California Exposition where the La Brea Tar Pits were opened to the public, displaying what were billed as the preserved bones and lifelike re-creations of Ice Age megafauna. Finding these bones intermingled with human artifacts in the dust, Ridgely knew he had something.

Rumors spread. Giant bones and weapons were sighted near Folsom, New Mexico, north of Clovis, almost at the Colorado border. The Eagle Scout did not relent. Aided by a prominent Clovis resident on the local Chamber of Commerce, he convinced archaeologists to investigate Blackwater Draw. By 1932, crews were digging in its banks, uncovering exactly what the Eagle Scout had described; megafauna and fluted points mixed together, a late Pleistocene free-for-all.

New Mexico is not necessarily the place that the Clovis archaeological complex should have been named after. More appropriate would have been Florida or Maryland, where fluted points have been found in greater abundance. Ninety percent of all documented fluted and Clovis-related points have been recorded east of the Mississippi, concentrated in lower Appalachia, the Gulf Coast, and the

Eastern Seaboard. This could, however, be an observer error. These are places of high contemporary populations, where artifacts are tilled up in farmers' fields, innumerable eyes scanning the ground. Modern population density, cultivated acreage, and intensity of archaeological research may have swayed the numbers. On the other hand, Western landscapes are often exposed and deflating, making it easier to find stone artifacts on the surface.

A geologist digging for stratigraphy on the High Plains of eastern Colorado told me that he'd found a Clovis-age artifact under twenty-seven feet of sediment. At the bottom of his trench he uncovered the cut and burned shoulder bone of a camel, meaning that what was on the surface in the Ice Age is now under twenty-seven feet of debris from rivers and wind. How much are we not seeing? This human-modified bone was found in what had been a shallow lake. The geologist gestured by rearing back his arm and throwing. He thought a camel had been cooked and eaten, and that some Clovis hunter sitting around idly afterward tossed the leftovers into the water.

Clovis being named after the town has more to do with the persistence of an Eagle Scout and a Chamber of Commerce than with science. I find this especially fitting. Science is useful, it fills in the blanks with precision, but history is ultimately more about stories and the unfolding of human whims.

Most of my family is from New Mexico. My stepmom and stepsister live in Roswell. My granddad died in a hotel in Albuquerque. My mom, the daughter of a telephone man, grew up in northern New Mexico, southern New Mexico, and the high desert in neighboring West Texas, where she was born. The surnames that make up my own history—Childs, Riegel, Bierschwale, and York—come from this mostly dry and windy country: children, nobles, beer-swillers, and tree people.

My granddad, now buried in the Roswell cemetery, used to drive me out onto the dry-grass plains where it seemed the wind always blew. We were archaeologists in our own right. Every few miles we'd see an abandoned house or a shed left empty for too many years, farmsteads gone under, turned to seed in the lesser dust bowls of the twentieth century. If the door was off its hinges or blown open, we'd go in.

I remember peaked roofs and wind-peeled doorjambs holding onto themselves against the sky. There was something eerie and enticing about these dwellings and their abandoned furniture, drawers that were empty and hard to pull out, wooden doorframes and porch beams that still carried invisible traces of the original atomic tests. The open sky of Chaves County, where most of my family lived, inhaled weather. Storm clouds on the horizon, the smell of moisture, the sound of rattling grass: These are what I remember. My granddad and I were curious about the last skim of life, finding a date on a scrap of newspaper crammed into a hole, part of a black-and-white photograph, a lone wooden crutch in the corner. I learned that there was more than one *now*. In these abandoned places, I could hear a knife chopping on a woodblock, a chair scooting out from the table, and words lasting late into the night, hushes and whispers turned to dust.

These are stories of my people, some of the memories that colonized this country. History books will show where we came from and how we spread, but they will not remember us, who we were, the reasons why we did what we did.

Why did they hunt the biggest animals on their landscape, when squirrels and rabbits have been shown to be three times more efficient as a source of calories? Notwithstanding Bruce Bradley's apocalypse hypothesis, it may have been a familiar human phenomenon rather than an end-of-the-world cult.

When I visited the village of Savoonga on St. Lawrence Island, I saw whale hunters, and their aluminum skiffs dragged onto tundra. They hunt the largest animals in their world, bowhead whales, even as these cold waters and island breeding grounds offered all the nutrition they needed in small packages easier to acquire than whales. In 2017, a sixteen-year-old hunter from St. Lawrence Island is credited with landing a fifty-seven-foot bowhead for the community of Gambell. "My Yup'ik name is *Agragiiq*," the young man reported. "The girls on top of the beach saw a whale, and they thought it was two of them, it was this bowhead whale. We went out and chased it for maybe an hour and a half; the other boats could have gotten it, but they never got close enough to strike. It came up right in front of us, and I struck it."

This same story might have been told by a young hunter in the late Pleistocene who delivered the first killing blow to a mammoth.

A whale hunt, or mammoth hunt by extension, is about more than calories. It is a cultural feature, part of being a people. Around Savoonga I saw skeletons and butchered remains of Arctic bowhead whales on the beaches where the Yup'ik community had, over decades, accumulated the remains of the largest animal they could hunt. Gary Haynes, an archaeologist at the University of Nevada in Reno and an American mammoth expert, wrote, "While the vast majority of anthropologists would shudder at the thought of approaching 1,000+ kg animals armed only with hand propelled weaponry, our hesitance cannot and should not be imposed on others. Recent forager-level hunters of elephant and whale are well aware of the potential risks involved in their predatory endeavors."

Mammoth hunts would have been historic, each with its place in local legend, some of the tales retold for generations. The sixteen-year-old whale hunter will not soon be forgotten. Hunts would have marked important moments in time, a spear stabbed in the ground, saying *here* and *now*. Back in the Ice Age, the whales walked on land, armed with tusks and horns, each footfall shaking the ground.

When Indigenous hunters return with a whale, the village rallies, everyone comes out. In Barrow, Alaska, long-handled knives and cutting spades known as cutting-in gear are brought to the site. The whaling captain's wife and her helpers hurry to serve hot drinks and boiled *maktak* to the butchering party before they head home to start cooking and preparing more food. When the butchering is done, usually within twenty-four hours, a flag is raised over the house of the boat captain who took the whale, and an announcement goes across the VHF that the feast is beginning, everyone should come. In this way, the hunt becomes about much more than resources. It is a way of defining a people, keeping them together at the edge of the world.

At Point Hope, Alaska, the captain who brought in a whale can be asked to hand over anything he owns; he is expected to fill the Thanksgiving and Christmas feast, giving everything away until his stores are empty, his family cooking for hundreds of people time and time again. Families might go broke, but the reward is a position in the community, the honor of taking a giant animal. That may be the reason people hunted megafauna in the first place.

The whale skeletons I saw were sinking into the ground, the larger idly etched with names and dates with pocket knives, or shot at for target practice. The biggest were dragged higher up from the water where their massive, sun-bleached bones became fixtures, a place to stop for a picnic, a giant vertebra serving as a bench where you could sit and watch the sea. They were part of a cultural, human landscape.

In New Mexico, we sat in the scrape and bluster of a windstorm, zinc-colored clouds whipping off the evaporated lake around us. We used our backs as shields and ate sausage, cheese, and mayonnaise, our bites gritty with gypsum. We'd left a camp in the dunes and were walking the alkali flats along a high edge of Ancestral Lake Otero. Storm clouds split into ghost fingers through the mountains and streamed over the dead lake.

We weren't far from Pendejo Cave, a day or two's walk to this rock-shelter in the side of a mesa that would have looked across the south end of the lake. Its name, *pendejo*, is a bitter curse in Spanish. The cave is a crummy little space, not quite as nice as Paisley Caves, a rough shelter where you could cook or sleep if you had to. Archaeologists thought they'd found a thirty-six-thousand-year-old human artifact here, as well as human skin prints in what had been damp sediment from the same age. Those dates, however, fell through, and scientists now say the oldest human presence at Pendejo appears in flakes and charcoal laid down about thirteen thousand years ago. This puts them in an early Clovis trajectory. A rough day in the field, giant cats out stalking the lake edge hungry for anything, bad weather . . . whatever drove them to this shelter kept them there long enough to leave rock chips and campfires.

On the lakebed, there was no shelter. Jordan was nervous about coming out this far. We were seeing more impact craters and military ephemera, shrapnel and wind-blasted pieces of equipment, though we hadn't yet seen the fence marking the edge of the missile range. Jordan kept asking who the trip leader was, who was making the decisions. He looked at me as he spoke, and said my name a number of times. I was trying to eat lunch and my eyeballs were sore. I told him we were all out here of our own volition, and it is perfectly legal to traverse the edge of a bombing range on public lands. Jordan said that was fine, he would do anything, he'd jump a fence on the range if necessary, but what, he asked, *were* we doing, besides sightseeing on a dead lake.

Beyond us was a dead zone, a scattered museum of war history, small parts of vehicles and sequential holes from long-ago strafing runs. Nick found a .50 caliber bullet. He turned it to show the butt, marked where the primer had been hit. The number "43" was imprinted at the edge. His eyes gleamed. "It was made in 1943," he said.

Meaning, it was from another age. Pre–atomic bomb, before depleted uranium was repurposed into tank-piercing projectiles. Of all of us, Nick was the best historian, having the strongest sense of what is at stake. In this bullet, he held a story from before the Atomic Age, before the world shifted. He knew what modern weapons meant. He'd pulled live bombs out of car trunks, puzzling over their many wires. A trail of blood led into the desert. A body shot through with shrapnel slumped against a steering wheel. He could smell the smoldering around him as he pondered booby traps within booby traps.

This may be why I seek the older stories. I flee to them, crawling into their caves for sanctuary. But I know better. It was no easier thirteen thousand years ago. We live within the same ranges of tension, going as far as we can without snapping regardless of the century or millennia. Bronze Age Ötzi was cut and wounded days before his death, and a flint arrowhead had been shot through his back into his shoulder; he probably bled to death where it cut an artery. Nine-thousand-year-old Kennewick Man in Washington State had a stone projectile embedded in his hip, the wound healed, bone grown over the artifact. A human burial on St. Lawrence Island in the Bering Sea revealed a skeleton pierced by fifteen barbed points made of bone, one skewered clean through the spinal column from behind. These were no easy deaths. If it wasn't being eaten alive by wild animals, it was someone throwing a spear at you, or maybe many spears. Nick knew this better than any of us.

I told the others to keep their eyes out for anything older than military artifacts, oddities sticking up from the playa, differently colored splotches of ground in a pattern. Mammoths wouldn't have left much and chances of finding anything in the course of a day's ramble across emptiness seemed slim. The tracks would be round expressions, like pancakes, or like badly cooked omelets. Ephemeral, washing away, they might be hardly more than discolorations raised an eighth of an inch from the ground.

When we found a trackway, I didn't know what I was seeing, or which way the mammoth was heading. One splotch, then two, a left-right pattern, some missing, some barely visible, seven prints in total. Jordan no longer cared how far out we were. We'd found something extraordinary.

This had been a Columbian mammoth, the tracks circular, decayed, and toeless. There would be no scientific report on the find. We'd never be able to find these again or explain where they were, compass bearings too vague on this expanse, no GPS to drop a waypoint. I walked alongside the tracks, and the mammoth rose up from the ground, its body filled in by my mind's eye. It didn't seem to notice me, it was focused ahead, tusks swaying back and forth as it traveled. It had hair, with rough brownish or gray skin visible underneath, but it was not woolly like its northern cousins, with which this species bred along the northern margins of its territory. This mammoth was a deep continental animal. It may have never seen its smaller woolly cousins.

With proboscideans, you can't tell by a simple track which way it was heading without seeing its toes. Nor can you tell how fast it was moving, elephants having only one gait, no sprint or gallop. All I could tell is that a mammoth had been here—*this* mammoth had been here. The tracks would have connected to trails coming in from the Jornada and from the mesas to the south. If they were anything like modern elephants, mammoths reused old routes, pounding them deeply into the ground. You would have seen their trails coming and going from all directions to this lake.

Mammoth tracks have also been found along a former lakeshore in Alberta, documented in fine-enough detail that researchers have been able to determine the population dynamics. Known as Wally's Beach, the site indicates a declining mammoth population at the end of the Pleistocene. The Alberta tracks show the same herd demographics of African elephants that have been battered by poaching and drought, the adults more numerous than juveniles. In a healthy African herd, juveniles represent 60 percent of individuals. A declining herd is 30 percent juveniles. That is what is seen at Wally's Beach. By the end of the Ice Age, mammoths were on their way out. There have been at least nine ice ages over the last seven hundred thousand years. The mammoths were weakened each time one ended, but they survived and rebounded to face the next go-round. This time they did not make it. For them, the last Ice Age was the end.

After the seventh track, the prints washed away. The mammoth kept going, walking until it was only a space in the air. I felt as if I were the one retreating, the one left behind as the mammoth lumbered into nothing.

Charlotte dropped to her knees about twenty feet from the tracks and pushed both hands firmly into the earth, giving an extra heave with her shoulders. She made two perfect handprints and scrawled the year of our arrival next to them, prints that would vanish in the next storm. This was our story. One geologic era later, mammoth hunters had found their quarry.

A friend of mine built a full-sized Columbian mammoth out of driftwood and brush, erecting it at the edge of the small desert town of Bluff, Utah, where he lives. His name is Joe Pachak, a local rock art specialist and a sculptor. A few years before building the thing, he announced that he'd seen what he thought might be mammoths on an ancient rock art panel. Along the San Juan River not far outside of town, he saw pecked into sandstone figures that appear to be animals with tusks, trunks, humped shoulders, and rounded heads, as well as other large animals, possibly a cloven-hoofed muskox, an animal that once lived in this part of the Southwest. The figures are obviously old, a style unfamiliar to archaeologists among the hundreds of other images and symbols pecked into a cliffside above the river, part of a palimpsest left over thousands of years. While some geologists argue that the rock face could not have held impressions for thirteen thousand years, archaeologists and rock art aficionados point out that the so-called mammoths are faded behind other images ascribed to early, pre-ceramic cultures. This would make the mammoths the oldest petroglyphs on the rock face.

Pachak had more than the rock art to go on for building his mammoth. A small Clovis hunting site was found on a ridge not much farther out of town. The Lime Ridge site looks to have been a quick stopover for a hunting party in the Ice Age, a site for repairing and prepping gear while watching the circle of the horizon from a subtle highpoint. In a region of soaring rock towers called Valley of the Gods, this flat-topped ridge is high enough that you can see all around from it, yet not so high that it breaks the horizon. When Pachak stood on it, he understood that Clovis hunters had chosen the location for the view, and also not to be seen.

Along with smaller flakes and blades, Lime Ridge produced a single fluted point made of a ghostly pale chert with a small, heart-shaped mineral inclusion in its face. The tip of the point had broken off and was found nearby. With no earlier sites found in the region, this may have been a late establishment in the deepest continental interior, a country of standing red rocks where horses, mammoths, and muskox grazed in grass that rippled like sea waves in a wind few people had ever known.

Lime Ridge looks down through a chain of gaps, where mammoths or anything else could have been driven. This would have brought prey down to a shallow, brushy creek, what would have been a cotton-wood bosque with ponds and marshes, now a dry, meandering wash at the foot of a dry, cliff-faced ridge. A mammoth leg bone was found in the next wash over, proof of their presence.

In a canyon seventy miles from Bluff, a shaded alcove known as Bechan Cave—pronounced *bu-shawn*, a Navajo word meaning "big shit"—has produced 255 cubic meters of mammoth dung with cannonball-sized boluses. Dung and mammoth bones found in the cave have provided six radiocarbon dates, ranging from 11,670 to 13,505 years old, which puts them within Clovis range. I spoke with a man named Cary Hightower who grew up running and culling wild horses in the canyon that leads up to Bechan Cave. He said that the canyon acted as a natural corral, and they used it to drive the horses into a place where they could be gelded and have their tails cropped. Hightower said that in the back of this bouldery, sandstone box, they corralled every ungulate that had been caught up together, not just the wild horses, but mule deer, bighorn sheep, sometimes pronghorn. He said, "And right above you is this cave full of mammoth dung."

Not an academic expert, but a man who knows the land and how people pass across it, Hightower believes that people in the Ice Age were driving mammoths off the flanks of the nearby Henry Mountains, herding them into this canyon, much the way he and his family did with wild horses. This cave may have been their last stand, their last local refuge.

Mammoth rock art, a small Clovis hunting camp nearby, and a cave blanketed with mammoth dung: That is a late Ice Age triad in southeast Utah. Other mammoth bones have been found in desert washes around the Four Corners, once a grassland among soaring red monuments. Pachak collaborated with a local artist, J. R. Lancaster, and the two gathered materials from desert washes and built scaffolding, erecting a full-scale mammoth on a metal frame.

The beast finally stood on all fours, about thirteen feet at the shoulder, bristling with wood, tusks made of two stripped cottonwood branches. On the Winter Solstice, the artists invited the surrounding community to an event where they burned the mammoth to the ground. The scene was festive and loud, heavy coats and warm hats on a chilly December eve, clear skies and stars out. Navajos and Utes came with drums and sang over the exhilarated gathering, some people dressed in battery-powered lights or masks, children running around, dogs trotting, wondering what all the excitement was about. This was the longest night of the year. Numerous rock art panels Pachak studied in the area had Winter Solstice alignments, so he thought it only fitting, a way of getting through the dark and cold, remembering a flicker of light at the other end. The flicker would be a burning mammoth.

A hush fell as an atlatl-thrower reared back and hurled a flaming projectile from thirty feet away into the beast's left flank, just behind the shoulder. The shot was perfect, the same place puncture wounds and broken-off spear tips have been found in proboscideans from the Ice Age. The projectile might have flown between ribs and shoulder bones, piercing the heart or lungs.

Flame quickly took hold as people roared. A second thrower came up and another atlatl dart flew. It struck inches away from the first, and the crowd erupted into shouts and howls. Drums immediately picked up speed as the mammoth was engulfed. Held up by its own girth and a metal framework, the sculpture did not fall right away. After half an hour, the head finally sagged. Two great arcs of cottonwood tusks slowly bowed toward the ground. The mammoth collapsed to its knees, then fell into a heap of glowing coals, stars showered with sparks.

Native American oral traditions frequently refer to giant, dangerous animals that ruled the land. In one Navajo story, these beasts gruesomely tore people apart and ate them. One of the monsters was a giant bird that snatched people and carried them away, a story known from here to South America. In Navajo tradition, it flew them over Shiprock in northwest New Mexico and dropped them to their deaths on the pointed top of the rock, which was said to be made of spearpoints sticking upward. They were pierced and hung, bleeding among skeletons, food for Monster Bird's chicks.

A pair of twin war gods came to kill all of these monsters in a story that can take years to tell. The slain bodies became part of the landscape. Monster Bird was felled by a magic arrow, but its chicks were allowed to live, becoming eagle and owl in one telling, helpers to people rather than enemies.

The Monster Slayer twins may have been Clovis, stories passed down through stories, hunters ridding the land of beasts too large and wild to live with. The magic arrow that killed Monster Bird could have been a point of a particular stone gathered from far away, reduced from both sides with bold strokes, and fluted up the middle.

There were ways to kill mammoths. The record of bone damage from human projectiles suggests an attack from around the shoulder blades, trying to drive the spear between ribs into its vitals, which involved jumping, thrusting, throwing, or otherwise getting in close. Spear tips from Siberia and North America are found embedded in the upper ribs or spinal area of mammoths, shoulder bones pierced with holes from projectiles.

Pressure upon pressure squeezed out the biggest animals, their final populations showing signs of disease, tuberculosis spreading among the giant bison, moving to other remaining big grazers. Columbian mammoths are thought to have endured longer in the well-watered hills, canyons, and mountain edges of the Southwest, while woolly mammoths survived in Wisconsin and points to the north. Mammoth and horse DNA found in soil samples along the banks of the Yukon River in central Alaska have been dated to between 10,500 and 7,500 years ago, indicating late surviving populations.

Without other pressures, humans alone could have cleared the field of megafauna. It has been estimated that if humans culled 3 percent of the mammoth population every year, North America could have been mostly mammoth-free within a few centuries. Some mammoths in the Far North escaped human encroachment, trapped on islands once part of the Bering land bridge. The last of these mammoths survived on St. Paul Island, largest of the Pribilofs, until fifty-six hundred years ago. Wrangell Island, in the Chukchi Sea off northwest Siberia, appears to have kept its mammoths the longest, the last of them gone four thousand years ago. The St. Paul mammoths evolved rapidly downward until they were the size of Percheron horses, and they appear to have succumbed to a lack of fresh water as the seas rose and their island dwindled, possibly dying in the same climatic event. The Wrangell Island mammoths were closer to full size and dropped to a population of about three hundred. They died off in what has been called a "genomic meltdown," mating within a closed loop, their hair losing its ability to insulate, proteins no longer adequately processed in the gut, glandular functions deteriorated, and sense of smell depleted. It would have been a rough and mangy demise.

Mainland mammoths became extinct many thousands of years before these island survivors. Battered by disease and a changing climate, they were now facing a coast-to-coast colonization by humans. When the last of them vanished is uncertain, between thirteen thousand and eleven thousand years ago, according to Bechan Cave in Utah and many other sites. Legends of hide and tusk would have finally faded from the Upper Midwest and down here in the Southwest.

This extinction event was the cry of a bullhorn, a piercing feedback loop. Native predators had to compete harder for prey reduced in size and number, while former giant herbivores died off from overhunting by every predator, humans included, atop punishing climate changes. It might have been a quadruple, quintuple whammy: comet impact, radiation from a coincidentally timed supernova, and a nitrogen collapse in Ice Age grasslands due to a shift in herbivore-ecosystem dynamics. The presence of *Sporormiella*, a soil fungus that appears specifically with large grazer dung, drops off rapidly at the end of the Ice Age. The pattern of this disappearing fungus indicates that megafauna populations fell first in the East and later in the West, as if cache-bearing hunters were traveling farther across the Great Plains to find their prey, finishing off what pockets still hung on.

Humans weren't the only invasive problem. *Bison antiquus* was also a factor, having come over the land bridge not long before people. They were non-natives, an arrival rarely good for home populations. Encroaching on the larger animals, spreading bovine tuberculosis — which has been detected in muskox, horses, and mammoths at the time—these bison may have been as much an extinction vector as we were.

As the Clovis culture started up, Ice Age foragers and megafauna hunters were entering the crash of a wave, North America warming faster than it had in more than ten thousand years, big, cold-adapted animals moving to whatever relic ecosystems they could as refuge. *Bison antiquus* shrank, eventually down to modern bison in size, adapting faster than most other megafauna.

When the big animals headed for refuges to wait out the coming interglacial, as they had done at the end of every ice age before this, they would have found people armed and ready. The site near Snowmass, Colorado, was one of these refuges. Hundreds of thousands of years of mastodon and mammoth bones and tusks have turned up, right up to the first inkling of humans, a possible placement of eleven stones and possible butchering marks forty thousand years old. After that, no more remains are found. The megafauna did not return. Snowmass is fifty miles north of a hilltop outside of Gunnison, Colorado, where late Ice Age hunters had a regional gathering between eleven and twelve thousand years ago, building a rock dwelling that

Muw-xar being driven into the underworld

overlooked the surrounding valleys where they lived year-round. They would have known their territory well. If a surviving mammoth herd plodded up the Roaring Fork Valley toward a refuge at Snowmass, following the picture of an alpine pond held in their atavistic memory, these hunters would have been there to greet them.

Low clouds and a blustery afternoon rolled into White Sands. The day was shadowless. We'd left camp with daypacks and were exploring north, finding fiberglass artifacts, trail markers from White Sands National Monument, a 275-square-mile park established in 1933. Trails, of course, were impossible to keep. They'd blow away within days, or the hour. Instead, posts had been driven in, each within line of sight of the next. We weren't on the trail, however, drifting as we explored the northern reaches of the monument, where dunes sheet toward the Jornada.

In the dunes, there is no reason to stay on one particular swell or another. It was like walking across the sea, our last day before we'd load up camp and head back to the vehicle. This is when Charlotte suggested a game. It would be like hide-and-seek, she said, but fiercer. I'd be a mammoth, maybe the last of them, and they'd be the hunters. I'd have a four-minute head start.

When I asked why me, Nick said, "What she's trying to say is that of all of us, you most resemble an old mammoth."

They were a couple of decades younger, Charlotte a global summit guide, Nick ex-military. Jordan, whose day job was welding, stands six-foot-four. Outrunning them would be impossible.

When Charlotte said "Go," I was already loping away, skirting a dune edge out of sight. Keeping my head low, coat unzipped, I started to pull off my wool scarf as I glanced back. Four minutes passed quickly. I saw their heads pop up, their raisin-shaped noggins obscured by blowing white sand.

I sought low passes into bottomlands below dune crests. At first they'd be jogging around, calling to each other. Half an hour passed and they were beginning to see this was real. I'd taken off. As an old mammoth, I hadn't lived this long by being stupid. My line had escaped the bloodbaths of kill sites where mothers and juveniles of my kind had been harvested. We had learned the range of a thrown weapon. It doesn't take long to figure out which animals to stay the hell away from. I doubled back through one basin and another, hugging the flanks of steep dune slopes.

If I knew these hunters, they would be scoping the crests, popping their heads up to look around. They were staying too close together, not covering enough ground. They thought I'd gone north when I'd gone south. Pale dunes blended into a cloudy sky, like looking into a bowl of oatmeal, which must have begun to weigh on their spirits. I put as much distance as I could between us. It took me an hour to lose them.

No proboscidean kills have been found in White Sands or the Jornada, but between Blackwater Draw in New Mexico and sites excavated in southern Arizona, it was happening. The San Pedro River flowing out of Mexico into southeast Arizona is peppered with Clovis sites with processed mammoths and bison, particularly the Lehner Mammoth Kill site, where at least a dozen young mammoths have been found with projectiles, butchering tools, camp fires, and the bones of bear, camel, tapir, bison, and horse. The pair of elephant-like gomphotheres recovered from the Fin del Mundo site in Sonora, Mexico, near the Sea of Cortez, were associated with fluted Clovis weaponry, marking the last recorded gomphotheres in the Americas, possibly hunted to extinction, at least given a shove over the edge.

I didn't want to lose them entirely. Even a mammoth would have appreciated the sport, an old itch being scratched. I climbed onto a dune crest and waited till one spotted me. I wanted them to know I'd skunked them. When their small figures fixated on me, I could see bewilderment in their stillness. The person on the far right, about the size of a rice grain, was Nick throwing up his arms, too far away to hear him shouting obscenities.

We hadn't talked about how this was supposed to end. With my death, I supposed. Or theirs. I had strong, eager, fervent people on my tail and now they were frustrated. They started toward me and I ducked away.

Most of the post-Clovis finds around White Sands—small overshot and fluting flakes, and pieces of projectiles from the later Folsom tradition—have been found at the far south end of the lake, south of Pendejo Cave below the Hueco Mountains. Their artifact distribution suggests a bit of a gathering, an area used with some frequency, while Mockingbird Gap, the Clovis site on the Jornada, was never returned to. These later Folsom hunters at the other end of the lake would have come into the dunes, tracking through these ghostly beach sands along a receding lakeshore. If a mammoth were here, they would have at least seen its tracks leading along a damp beach where they may have marveled that such a beast still existed.

Hunters of both Folsom and Clovis age are considered "communal hunters" working together in a variety of well-defined roles to take down shockingly large animals. If my friends didn't have a plan before, they did now. Spread into a fan, Jordan was on one side, Charlotte on the other, Nick in the middle. It didn't matter how far I ran, they would close in like a net.

I tried to keep ahead of them, but I was tired. After fifteen minutes of running, I slowed to a walk. I pulled out my journal and paused to write, waiting for my fate.

My species had lived a good, long while, that much I concluded. The first mammoths emerged in Africa about five million years ago, and moved into Europe two million years later, spreading from there across Asia. About 1.5 million years ago, the giant steppe mammoth *Mammuthus trogontherii*, one of the largest proboscideans of all time, crossed an early Bering land bridge and entered this hemisphere, where it traveled south and evolved into the Columbian mammoth. I thought of myself, the mammoth, as the quintessential American, here long enough that I not only came from the ground, I *was* the ground. My kind would eventually be forged in the crucible of Clovis, a brave and terrible people. Now who were they? Folsom hunters, a new kind of technology, miniature Clovis. It seemed there were more of them all the time, guarding the passes and lake edges in an age when browse was becoming harder to find.

The Mammoth Steppe was in the process of collapsing. Without big grazers, biotic turnover slowed and grasses turned to shrubs and tundra, no longer sequestering as much CO_2, turning off one of the hemisphere's cooling engines. The Ice Age was shutting down. Mammoths weren't getting out alive, not this time. People would have known what was happening. It might have eventually dawned on them that a slice of their world was disappearing. In the dimming Pleistocene, they would have found big game farther and farther apart. If they saw the end of mammoths coming, would they have stopped killing them?

When Nick found me on a dune pan, it was somewhere between

eleven thousand and thirteen thousand years ago, and the wind was stinging our faces, the chill of December settled in. I looked up from my journal and from half a football field away we saw each other.

I jammed the journal back into my daypack, pen into my pocket, turned, and ran.

Nick cursed and broke into a sprint.

He was a javelin, his momentum unbreakable. At the last second, I dodged right when he meant left, and he flailed. One flying arm snatched my daypack. I slipped out of its straps and the pack flew free. As I ran off, he came to one knee panting. He threw my pack out of the way and started up again.

Jordan must have been watching, waiting. He sprang from behind a low dune. Thirty yards away and coming straight for me, he roared, reaching out with orangutan arms. I could see his teeth. I dove away and he tumbled across the pan, landing on his knees.

I ran away, yelling, "Fucking bison hunters!"

Charlotte wasn't around. Unbeknownst to us, she had drifted to the far side of the hunt. She'd taken out her contact lenses in the blowing grit and was near the monument trail with markers spiked into the sand. As she would later tell us, when she saw three figures walking over a high dune, she sneaked up below them, having pulled up one of the markers to use as a spear. Catching a mammoth and two human hunters by surprise, she was about to count serious coup.

Charlotte planned to spring cat-like, a lone hunter, a feral creature. Shying her way up the side of a dune, it would be a dash, a hundred feet or more with the pointed end of the marker aimed at us. She'd throw it and it would spear into the sand between us. She readied herself, and jumped. The instant her head cleared the sand crest, she clambered backward and disappeared.

It was not us; it was a family of three, a mother, father, and child bundled against the wind, out hiking for the day. Horrified, Charlotte shrank down the dune, narrowly unseen.

At about the same time, I was out of energy. I was limping. Somewhere in the fray, I had pulled a muscle in my groin. A dune slope at my back was all I could do. I staggered with hands on my knees and looked for any other exit, but the hunters were already here.

I turned with my ivory tusks, their great arcs towering over two men approaching with wild and wary eyes. They appeared drained but undaunted. They knew what was about to happen. From ten feet away I locked eyes with both of them. No words. Just bodies. They came in slowly, they had all the time they needed, breathing hard from the run as they drew to within arm's reach.

I faked one way and another, one last chance to slip the trap. Nick dove and hit me in the knees, arms wrapped around my legs. Jordan was a quarter-second behind him, his full weight plowing into my torso. I felt my tailbone hit the ground. The back of my skull made contact with sand. Breath slammed out of me. The hunters rolled off where they lay panting on the side of the dune. It was done. The mammoth was dead.

10

AMERICAN BABYLON

12,800 TO 11,800 YEARS AGO

When the front doors opened at Stewart Hall on the St. Cloud State University campus in central Minnesota, the ceiling heater kicked on and blew the cold back outside. It felt like stepping into an airlock. The Upper Midwest had been bottoming out at 50 degrees below zero Fahrenheit, highs across the region not reaching above zero for weeks. NASA images of North America that January in 2014 looked like the heart of the Ice Age, the top half of Northern Hemisphere blank white, Great Lakes mostly frozen. It was one of the coldest winters on record. The polar vortex, which sits on the top of the globe like a halo, had broken loose and hemorrhaged through an unstable jet stream. Hoops and gears of climate were wobbling and springing apart, reliable seasonal patterns breaking down, as Arctic air fell across the Northern Hemisphere like a hood.

I stamped off the snow as I came in. Night had fallen, classes were over, and the halls of the building were quiet. The last few nights I'd been sleeping outside, exploring late Ice Age conditions, a prolonged cold spell that came like a punch to end a geologic era, and I'd never been so cold in my life. As I walked down the hall, I left a leaden trail in the air from the shell of my coat. I pulled off my mitts, my hat, unzipping, letting in the warmth as I looked for numbers and signs on the walls.

I was here to meet archaeologist Mark Muñiz, who had been focusing on one island amid the hundreds of lakes in the Boundary Waters of northern Minnesota. The ground at his isolated site was covered with so many late Ice Age tools and pieces of debitage—stone flakes and debris—that he'd told me you couldn't walk without stepping on them. This was the Knife Lake Quarry Complex, a twelve-thousand-year-old gathering and processing industry around exposures of a fine-grained Precambrian siltstone ranging in color from white to gray to chalkboard black. The rock comes off in tabular pieces, perfect for knapping into tools and projectiles. Knife Lake siltstone comes solely from northeast Minnesota and was carried hundreds of miles to the south, one of the exotic toolstones commonly found in the Upper Mississippi River Valley of Wisconsin and Illinois.

A few months earlier, I had hatched a plan with Muñiz to access Knife Lake in the winter by dogsled. This would allow a perspective on the Younger Dryas, a cooling anomaly that kicked in 12,800 years ago and changed the course of the first people. Muñiz had gone to the site only in warmer months when waterways were open, accessing Knife Lake by canoes with gear and crew for the dig. I wanted to see the site in its cold phase. We picked the wrong winter to attempt our trek. By the start of 2014, the outfitter supplying the dogs and sleds said the snow was too deep and the nights were too cold. The polar vortex had unraveled and was frosting the continent down to Texas, 20 degrees in Houston. The outfitter wasn't sending anyone out at 50 below zero, neither dog nor man. I opted instead to hike across a corner of nearby Lake Superior, stomping ahead in moonboots and dragging a toboggan loaded with gear on a frozen, empty bay south of the Apostle Islands. Muñiz wisely declined, and for a few days I was adrift in searing cold.

The name "Younger Dryas" comes from the diminutive, hardy *Dryas octopetala* flower, whose presence in soil horizons tells how far south in Europe polar conditions reached. Frigid climates are marked by these ground-hugging, white-petaled flowers. *Younger* refers to the fact that there was also an *Older* Dryas, a cooling spell that occurred around 14,000 years ago. It lasted a century. The Younger Dryas lasted a thousand years.

The cooling event I witnessed from Lake Superior was my own Younger Dryas in miniature. It began in late 2013, with a warming episode in Asia. India grew hotter than usual and bounced a dome of warm air off the Tibetan Plateau into the stratosphere. The jet stream, a governor of northern climates, was severed, upsetting the polar vortex over the Arctic. The vortex reversed itself from its original counter-clockwise movement and began turning clockwise instead. It's a worrisome change when a planet flips one of its primary circulation patterns. Cold air pushed through the weakened jet stream and escaped southward, creating climatic mayhem at the top of the planet. Lake Superior froze more solidly than it had in almost five decades. Ice would still be floating in the lake through May.

The Younger Dryas began 12,800 years ago, essentially restarting the Ice Age, returning to polar conditions that had been phasing out over the previous several thousand years. Within a decade the Northern hemisphere went from warming straight back to deep and windy cold. The main cause was likely a thermohaline issue. Floods of fresh, cold meltwater began coming off North America, entering the slowly warming Atlantic Gulf Stream, which, like the jet stream, is a climate governor for the Northern Hemisphere. Saltwater sank under the inundation of freshwater, and the surface began to freeze solid, reversing the belt-like circulation of the Gulf Stream. Rather than shuttling warm equatorial waters north as it had for six thousand years, the Gulf Stream now began pulling cold North Atlantic waters south. Icebergs began showing up off the Atlantic coast of Florida.

The most likely trigger was flooding from a giant lake that had built up against the previously shrinking Laurentide Ice Sheet. The freshwater burst happened as the retreating ice collapsed, allowing a lake ten times the size of Lake Superior to pour out and reach the sea. Another large meltwater discharge came from a shrinking ice dome that spilled at once into the Arctic Ocean via the McKenzie River. These floods might have been enough to turn the Ice Age engine back on.

This new, thousand-year-long cold spell was devastating to megafauna, sending the last big animals out of the Ice Age and into extinction. But humans took the change and ran with it, adapting to the new conditions and breaking into technologically distinct groups. Stemmed points outpaced fluted, becoming floristic. Styles diversified: Great Basin points from mid-continent; straight-stemmed Lake Mojave projectiles along the California/Nevada border; squat-bottom Windust stemmed points from the Pacific Northwest. Cougar Mountain points from the upper Great Basin held on to the oldest traits, still resembling those found in Paisley Caves from a few thousand years earlier.

Clovis, as a continental manifestation, ended with the start of the Younger Dryas, and spread into numerous regional offshoots. Weapons mostly became smaller and more intricately made to match the new prey size. By becoming even mildly isolated, fluted technology evolved rapidly, styles dividing into many. Long, fluted, almost absurdly Clovis-like projectiles with delicate owl-ears at the bottom were manufactured in lower Appalachia and throughout the Tennessee River basin. Shorter, thumb-sized Folsom points spread across the High Plains. The biblical story of Babylon comes to mind, a mythic time when all people spoke the same language. They gathered to build a city and a great tower, which God promptly destroyed. At that moment, the people could not understand each other, as if suddenly speaking babble. They had been granted mutually unintelligible languages. Forced into diversity, they scattered from each other into tribes and clans.

In late Ice Age America, technologies diverged, and probably dialects and languages along with them. Once few, people became many. Smaller fluted points appeared in western Alaska near the edge of the dwindling land bridge about twelve thousand years ago, backwashing into Asia, where a fluted point was found at the Uptar site in Kamchatka. Fluted points also appeared in caves in southern Chile about eleven thousand years ago. Despite the cold, or maybe because of it, styles were flourishing. It was a hard, swift break on the pool table of America, the Younger Dryas the cue ball.

Before meeting with Muñiz I had set up a simple camp on the white expanse of Lake Superior, on the north shore of Wisconsin, no tent, just a bag, bivvy, and the toboggan I hauled behind me onto the lake's surface. The only heat I carried was my body, which was enough till about midnight the first night between Ashland and Washburn. After that, it was a struggle to keep rolling and cupping whatever warmth I could to avoid frostbite. I'd gone out for a taste of full glacial conditions, listening to lake ice splinter and buckle beneath me, the lake's surface crushing itself at 30 degrees below zero. What I found in this slice of the Younger Dryas was mostly misery, which might have been the point. After getting no sleep, crawling out after sunrise from zippers frozen solid, and trying to start my stove with a lighter in clear frigid sunlight—the gas too cold to ignite—I became convinced that the people living against ancestral great lakes and the ice sheet in the Younger Dryas must have been seasonal migrants from the South, coming only in warmer months. I thought I should tell Muñiz this. They were in Florida warming their toes in the winter, I was sure of it.

But I was not Paleolithic. I was a very cold person from the Holocene. Instead of hauling a toboggan of gear, I would have been moving rocks, taking prizes from Knife Lake out to where I could trade or fashion them into weapons. Archaeologist John Lambert at UC Davis studied private collections of Paleo projectiles found on the prairie of eastern Wisconsin and traced the rocks back to their sources. He found that while some late Pleistocene artifacts were made with local raw materials, the majority came from "high-quality, exotic toolstones transported 300–500 km," many coming from around the Great Lakes and the top of Minnesota. Lambert noted that toolkit composition in the region reflects a reliance on long-distance movement, and that the rocks were transported along a north/south axis. That fit my personal hypothesis that they were wintering in Florida. Or in Tennessee where projectiles similar to those around the Great Lakes were being made, Barnes or Cumberland points. Muñiz didn't picture them ranging as far as I envisioned, but we agreed: People did not stop moving when the Ice Age turned back on.

I looked into an office and knocked on the doorjamb. Muñiz turned from his desk and welcomed me in. A stout, broad-shouldered man, probably good in a brawl, he was in his forties, and would have done fine alongside me, freezing his ass off on the face of Lake Superior. I piled my coats in the corner because they wouldn't fit on his door hook. He asked where I finally ended up. When I told him out near the Apostle Islands, he laughed and said he'd made the right decision.

Muñiz had his desktop open, layers of student work, grants, documents of survival for a university archaeologist. The desk had books and papers stacked, illustrations of preforms and points. He told me that during a winter like this, you think about being anywhere but here. Florida, maybe, cottonmouths, alligators, and all.

Muñiz had cut his teeth on Paleolithic sites along the Aucilla River.

He spoke of hot summer days when he'd dive into a hole and sink to the bottom, cooler and cooler as the weight belt carried him down. He held his arms out in front of him, showing how he'd descend as if in space. Landing on the bottom in darkness, breathing out of a regulator, he would wait several minutes for the sediment of his arrival to settle. He'd listen to his breath and bubbles rising, no light, not even a faint sun. Flicking on a thousand-watt spotlight, he described how megafauna bones would appear, his spotlight bouncing off them through the tannic water like gold.

"You ever see *Pulp Fiction*," he said, "where they open the briefcase and they're looking into it?"

I knew the scene, and you don't actually see what's in the briefcase, but whenever they open it, you see the glow on their faces, their eyes frozen for a moment over something mysterious, entrancing.

He said, "That's what it was like."

Now he's researching cold-period people who acquired stone and fashioned it into tools and transportable rock at the frigid, newly expanding edge of the Laurentide Ice Sheet.

While the coasts were shrinking from sea-level rise, fresh earth was exposed by glacial retreat over the previous several thousand years. At its greatest extent, the Laurentide reached across most of Minnesota, and covered all of nearby Canada. It had melted and collapsed more than a hundred miles back before the Younger Dryas. Then it slowly began to grow again, looming like a dome across Ontario, spilling icebergs into the chalky expanse of ancestral Lake Duluth.

French fur traders in the eighteenth century called the lake around Muñiz's site *Lac des Couteaux,* Lake of Knives. They got this from Ojibwa people, who called it *Mookomann Zaaga'igan,* which means "knife-rock lake," a place where knives could be made from rock. Radiocarbon dates from the site put its original use at twelve thousand years ago. Muñiz told me that these people might have been recent arrivals from the North country, part of a movement coming down the ice-free corridor, people who knew something about cold, perhaps preferred it. The Younger Dryas may have been their time to flourish.

He said the people who came here for toolstone were firmly in touch with the larger geography, moving their resource out of Knife Lake two to three hundred miles into surrounding regions, while weapons found in the area came from different continental traditions. Long, lanceolate points appeared at the quarry site, Clovis-like though intentionally not fluted, and yet also with a stemmed design, showing a mix of techniques at the dawn of the Holocene. Similar points are found across the Midwest and into the North Plains and the newly exposed land of Canada. They are similar, in turn, to points found at the Mesa Site north of the Brooks Range in Alaska, a sign of a later group possibly coming down through the ice-free corridor, connecting northern Alaska with the American interior.

Muñiz does not see a well-populated land twelve thousand years ago. Instead, he envisions remote lake edges, hundreds of miles between one band of hunter-gatherers and another. The Younger Dryas people who arrived at the shores of Lake Duluth may have been caribou and muskox hunters tracking edges of ice and lakes, finding this shiny source of siltstone out here like a gem. He showed me maps of the original shorelines, the quarry out on a lake in what must have been the middle of nowhere, tundra and permafrost, winters frozen solid.

"There are plenty of animals elsewhere, plenty of other good stone," Muñiz said. "To me, their arrival here has got to be exploration. But it's hard to build a scientific interpretation when it's human nature."

Muñiz took me to the lab through empty hallways, unlocking doors, the building humming to itself with one perpetual breath, air ducts rattling. Pulling out drawers, bags, and boxes, he displayed samples of stones he and his students collected, hundreds upon hundreds of large flakes, unfinished blanks, bifacial knives, end scrapers, adzes, and preforms. He passed one after another to me, pressing them into my hands. The rocks told him that this was a prep zone, where flakes were knocked off, mass reduced, preforms made ready to go. A supply of roughed-out bifaces, enough to carry and make the trip worth it, could easily be made in a few hours and hauled away.

He'd watch how his students on the dig and survey crew set up their tents around the quarry, clustered together for the most part, some set off a ways. This, he imagined, was the way people did it twelve thousand years ago, settling themselves on the land, leaving fire hearths and chipping areas, temporary camps for seasonal cycles.

He showed me what he thought was an overshot flake, a memory of Clovis. Some of the almost finished projectiles he'd picked up were long and intimidating. He called them "shock-and-awe weapons."

I asked him about the fall of Clovis at the start of the Younger Dryas.

"I wouldn't call it a fall," he said. "More like, diversified."

Whether Clovis was a people, a technology, or a cult, it had to adapt or perish. The focus on mammoths had to change. Now bison, muskox, caribou, and elk were the largest animals. Muñiz doesn't see people dying off, but adapting, taking on new appearances and locations in the archaeological record.

"You know the difference between Clovis and us?" Muñiz said. "When the climate changed and things got rough, Clovis had somewhere to go."

Before visiting Muñiz, I spent only one night on the lake. As temperatures dropped to around 30 below zero, where it doesn't seem to matter if it's Celsius or Fahrenheit, Paleolithic or Holocene, I kept imagining dire wolves or the tusked hulk of a proboscidean appearing on the half light of the ice. If any were still alive after the Younger Dryas, this might have been the place to see them. A mastodon found in the southern Great Lakes region dated to between ten and eleven thousand years old, one of the final survivors before extinction, a last, lost individual.

After that sleepless night fending off frostbite, I marched inland to camp in a snow shelter that I'd helped build a few days earlier with students from a college in Ashland, Wisconsin. This was my version of adaptation. I wasn't going to try another night on the lake under these collapsed polar vortex conditions.

The shelter I helped construct was a *quinzee*, a circular room with a domed interior dug out of snow. It could sleep five or six. The entrance was a hole as wide as your hips, like slipping into an underground burrow. I crawled into it that night, pushing gear ahead of me, anything that I didn't want to freeze solid outside. My down sleeping bag was like an animal carcass, and I sat in it wearing my boots. It felt as if I were deep inside the earth, no sound, no sense of direction.

Students had left tea candles and a thermometer. I lit the candles in a circle, my anorak hood pulled back, then the layers of warm caps and mitts off, fingers feeling the air. My breath curled around my head, fogging the ceiling. Inside temperature rose from minus 27, leveling off at 10 below, which felt comfortable enough that I unzipped my layers of jackets.

As I entered my bag for the night, one candle remained lit. Its wick drew up the last clear wax, a soft light cast across the snow-packed floor and walls, ceiling made of scrape marks from snow shovels. I lay on my back reading the marks like cuneiform.

The candle sputtered, gave a few winks, and the chamber fell into darkness. No stars, no sky, no Earth. I could hear my breath. How many nights, I wondered, did they sleep this way, going under for winter, hiding from the wild and toothy exterior? I thought of the huts and caves, the shallow shelters of Bluefish and Paisley, the rough hole of Pendejo Cave in the side of Otero Mesa outside of White Sands.

This is where, at the beginning of time, Raven strolled along a shoreline and came upon a mysterious clamshell. It was closed tightly. He put his ear to the shell and inside he could hear breathing.

THE PARTY AT THE BEGINNING
OF THE WORLD

11,000 YEARS AGO

Fires and spotlights lit the distant sky, the night beginning to glow. If we'd been closer, we could have heard drums and seen soaring flames, but we were still days away from the city.

In the morning, I opened my eyes to clear, clean light. It was the sunlight of the Holocene, the glacial age giving up its ghost. I had spent the night on the floor of this waterless lake in the desert of northwest Nevada—Lake Lahontan, one of the largest water bodies on the continent eleven thousand years ago. It once covered parts of Nevada, California, and Oregon, back when glaciers were melting, rainclouds gathering, and the Younger Dryas was over.

The lake has become a network of dusty inlets, peninsulas, and pans across the Black Rock Desert. We'd camped between ragged mountains in a small, protected bay, its floor as smooth as a skating rink. Other sleeping bags were out, some occupied, some not. Six in total. Backpacks were strewn around, signs of travel. Not much was needed on a summer night here, a limp sack of down to crawl into, the slightest cool of evening already washed away by sunrise.

I got up and put on a thin sarong, a sun shirt, hat, and walked off barefoot. We brought the lightest gear, no stoves, no coffee. Bullets lay on the ground from people shooting at nothing, target practice on the horizon. None of the bullets, .22s and .45s, were marred or misshapen. They'd been fired and flew across the pan until they lost momentum, skidding on soft sediment as if landing in a pillow. I collected a handful, like finding marbles, then cast them back out.

At the close of the Ice Age, at least ten distinct groups in the Americas can be identified through their stone tool and weapon traditions. Archaeologist Michael Faught, currently mapping the floor of the Gulf of Mexico for early human inhabitants, identifies these ten lithic groups as *Nenana, Denali, Fluted Point, Western Stemmed Point, Flake Tool, Flake and Biface Tool, Fishtail Point, Paijan, Uruguai,* and *Paranaibo.* Paleolithic America was never a monolithic culture, it was always many cultures, especially near the end, as people divided and came together. Faught wrote that skeletal samples have revealed that "later Holocene crania in North America have similarities with northeast Asian samples, but early Holocene remains in western North America, Middle America, and South America do not. These earlier samples have more cranio-facial similarities with Australian, African, or Melanesian populations. The salient point is that diversity is indicated."

The couple in our group, a fiddle player from Bali named Etha Widiyanto and her lanky Grand Canyon river guide boyfriend, Thomas Patrick O'Hara, were curled up like a pair of caterpillars in their joined sleeping bags. The rest of us were single and up for the day. John Tveten, an ER doc who used to work on the Hopi Reservation, played mandolin as he strolled into sunrise. Q, the same guy who'd joined me on the Harding Icefield, a brawny, Alaskan-born photographer, was out with a tripod, time-lapsing shadows as the first sun shattered the mountains. Timmy O'Neill, a climber friend, a stand-up comedian and drummer in a rock band, was already shirtless and juggling, wandering across the playa throwing balls in the air. We were a traveling minstrel show. Timmy would play the drums on whatever he could find, water bottles, knife blades, or rocks. He'd brought a tambourine with him. The doctor had his mandolin, Etha played fiddle, O'Hara a guitar. Q sang sometimes between filming, and I'd drop in to chant a story to their howls and drumbeats, O'Hara singing behind me with so much force that his neck veins popped out, performing to nothing but desert.

We were taking our show to the city being built beyond the arms and peaks of Lahontan, the site of Burning Man. Its glow brightened as we approached night by night. Spotlights fanned from behind layers of mountains still six days away. The glow was a beacon. It said, *Come.*

Burning Man, our camp, and everything at eye level this morning would have been nine hundred feet below the waves, a ghost circus in the murky depths of Lake Lahontan. When early people were here, they occupied the surrounding shores and palisades, looking across water and islands. One type of artifact they left, unique to lakes country, is the "lunate crescent," a kind of weapon, scraper, or cutting tool found among marshes, estuaries, and islands from the Intermountain West to the California coast. Also known as a "butterfly" or "winged" crescent, this stone weapon or tool—no one is sure how it was used—appeared around twelve thousand years ago during the Younger Dryas. Based on where it's found around water, the artifact is believed to have been associated with bird hunting or skinning, though one crescent tool produced signs of pronghorn blood, and I've thought it could have as easily been used on fish.

These stone crescents are typically thin, flat, and the length of your finger, as symmetrical as a boomerang. They often have grinding around the midline to facilitate hafting onto a wooden shaft. It may be coincidental, but their distribution follows flyways for tundra swan, greater white-fronted goose, snow goose, and Ross's goose, lending credence to the crescent being an implement designed for hunting birds. Whatever they were used for, they were unique to this part of the world, possibly a regional clan of bird hunters strung along lakes at the end of the Ice Age in the American West.

As I strolled across the bottom of Lahontan, I came upon four ibises. They stood at attention with stilt legs and long shadows. When they stiffened, I stopped.

These were waterbirds, their element conspicuously absent. They seemed shy and, for all their grace, awkward. They may have landed here like us, curling up instinctively in the cup of a long-gone bay, the only place they could find that looked like water. This was one of the last great Western lakes to evaporate. One bird took a step and another, and they stilled, their bodies black against the rising sun, the playa turning into light around them.

The numbers of birds coming through here now are infinitesimal compared to the Ice Age. These pans are still used as flyways, and what few lakes remain are busy wintering grounds. During the late

Lunate crescent and ibises

Pleistocene and the first few thousand years of the Holocene, millions of waterbirds are thought to have leapfrogged these lakes every season. The lunate crescent followed them, and when most of the lakes finally dried up and flyways scattered eight thousand years ago, the lunate crescent went out of use.

The ibises and I watched each other like unexpected strangers, not sure how to relate. Beyond the playa, mountains rose like nunataks, summits pierced through ice, through the chalky sediments of the lake bottom. I took a wide detour around the birds, not wanting to make their journey any more difficult. I wished for them actual lakes, wherever they could be found in this drying age.

Most of the continent was hit hard by the chill of the Younger Dryas at the Ice Age's end, but not every place suffered. Lake Lahontan

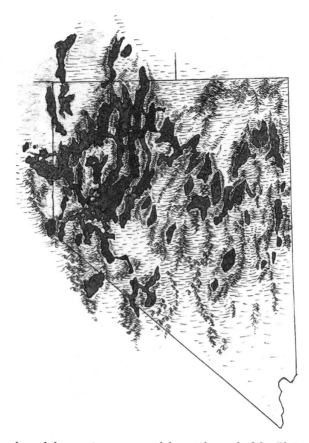

Nevada and the maximum extent lakes at the end of the Pleistocene

and its many surrounding sister lakes reached their biotic climax in the thousand-year-long cold spell. This was ideal country, plenty of water, mid-latitude warmth, far enough from the glaciated Rockies and the Coastal Ranges not to be hemmed in by ice, its lakeshores rich in marshes and fishing. The geology has a wealth of jaspers and cherts, and good obsidian sources not far away for making tools and weapons. Terraces are plumed in tufa formations from freshwater springs that would have poured into the lake like fountains. It was yet another Eden in the long history of discoveries.

The Black Rock Desert has kept summers of field schools going with researchers, grads, diggers, and mappers. This is one of the Paleolithic center-points of North America, the Intermountain West chained with great, dry lakes, from Paisley Caves in Oregon to White

Sands in New Mexico. Lake Lahontan was the largest with 8,500 square miles of surface area, nine hundred feet deep at its high-stand. The mountains around our camp would have been islands, their summits reflected across the water. We were in the bottom, the muddy floor now hard-packed and dry.

People who lived at this lake were big- and small-game hunters, shore fishers, eaters of marsh grass and *Apiacea* roots—the same food the first people in Paisley Caves had eaten. They used spear-points and lunate crescents, some of their weapons identical to those found in Paisley Caves, meaning the stemmed-tool tradition was alive and well through the end of the Ice Age. Perhaps by eleven thousand years ago, people had forgotten where they came from, their stemmed weaponry no longer from the Pacific Rim, now from here.

The giant predators were gone. Grizzlies, mountain lions, and timber wolves were the largest carnivores in their world. American lions, sabertooths, dire wolves, and short-faced bears were legends from a world before, when monsters ruled the earth and humans fought for purchase.

Faces dusty, packs heavy with water, we wound along lakeshores as dust devils careened across the playa, columns of alkali twisting into a blank sky. I'd come to think of these whirlwinds as spirits, or ghosts—whatever word you might use. Even the tiny ones, the quick whips of air twisting in tight circles, are comments, words, recollections. The big ones thundered across the playa, large as houses, feeding the atmosphere with dust.

It is a ghostly land when you pay attention. There are places you can't walk without a shiver trickling up your spine. Have you gone across the Navajo Reservation with its red butte towers and spikes of eroded volcanoes? Human bones wash out of the cutbanks with pockets of beads and broken pottery. Everywhere you go, there are drums in the ground.

Etched into the mountains, the terraces of Lahontan looked as if a dam had burst and drained out. Hundreds of feet above the playa, we walked through cliff edges and steep, crumbling slopes, contouring along these abandoned watermarks. We found occasional sites, old beaches flaked with butterscotch-colored stone people had brought in and chipped.

Midday, we retreated under our umbrellas, except for Etha, who huddled beneath a light shawl, and O'Hara, weathered and middle-aged, sprawled in the sun and cooking like a lizard. Wind whipped across us as if exhaling from a furnace. Nobody had played music in a while. In the evenings, people would break out instruments, and sometimes during a water stop in the middle of the day. This afternoon, instruments stayed in their cases as wind streamed across the highest shore of the lake.

The mountains around us were fissured with steep, rocky draws. Timmy, the climber, walked under his umbrella in the mid-afternoon swelter, looking for artifacts, while the rest of us lounged in our half-circles of shade, umbrellas propped up against the wind. He looked like an archaeologist on methamphetamines, bending down, picking up, putting back, swiftly moving to the next. He's always like this. I felt like a hobo around him, while he bobbed and flew like a hummingbird. Now he was after artifacts, the next shiny thing. I described the shapes I was seeking, and within half an hour he'd found four late-stage preforms probably from the Paleoarchaic, a time bracket of a few thousand years, anywhere between the end of the Pleistocene and the beginning of the Holocene. It was a time of more rains and fewer snows than the Ice Age. Those who had survived the thousand-year winter of the Younger Dryas were now in the right position for the coming age.

Once he heard he might find a Paleolithic spearpoint, Timmy started covering ground like a ferret. He walked in and out of draws, lean body on the move, picking up one piece, then another, bringing them back to show us under our umbrellas. He was finding more preforms, one of the original human travel tools. One was a cloudy white chert, two others were the local butterscotch. They were broken, tumbled down a draw, but they got me up and I started walking, looking, reaching down, finding the wind-whittled edge of a scraper, and a core smaller than a fist from which knife blades had been cleaved. Soon umbrellas were stowed in packs and the six of us were following a trail of encampments. Every hundred feet or so was a new concentration, knapping circles and hunt-ready spike camps. They'd positioned themselves on a palisade over the lake. I thought they were hunting deer, bison, or elk, prey large enough to warrant a projectile made from a hand-sized preform.

Timmy wanted a final-stage Paleolithic projectile. I told him about the stemmed lance known as a haskett. One was found along a dry

Utah lakebed with the residue of proboscidean protein in its micro-scopic cracks. Timmy said he'd find one of those, too.

What he found was a sunset camp looking across a pale gulf of lakebed. You could have run off the edge with a hang glider and taken flight with a thousand feet to spare. The band broke out. Etha yawed on her fiddle as O'Hara and the doctor followed her up with their guitars. Instead of drumming, Timmy moved up and down the draws in evening light. He still wanted to find one of these great, ancient points. He wouldn't, though. In decades of searching, I never have.

The Ice Age was done, and people were still going. In the transi-tion between geologic ages, they banded into groups, forming the first large aggregations, gathering at places like Lake Lahontan and its cadre of intermountain waterbodies. Imagine fires on these shores, lights flickering far away, people letting each other know they were here. A rock art panel on a lake arm south of here had been attended to continuously from 14,800 years ago to the close of the Pleistocene, people coming back to it over and over for thousands of years.

Another, more condensed gathering place is in the Northeast, where large camps collected around caribou drives. The largest site is Bull Brook along the lower Ipswich River on the Atlantic edge of Massachusetts. Twice the size of any other gathering in the North-east, Bull Brook was a hunting locus in coastal tundra and woods. It appears to have been used just once, eleven thousand years ago, heavy in fluted weaponry smaller than Clovis, more fitting for the prey size of the age.

Bull Brook has been destroyed by development, the site lost since the 1950s. Before bulldozers moved in, an archaeological inventory mapped out the distribution of artifacts. Thirty-six discrete sites or encampments were found in a circle or semi-circle around a central area, with far fewer finds in the middle, a sort of plaza within a social aggregation. Artifacts found in the outer and inner rings are notably different and appear to be hierarchically organized. Fluted point pro-duction is concentrated in the innermost ring, while spurred stone gravers used for bone industries and punching through skins, as well as hide-working scrapers, are more common around the larger, outer ring. The site was well ordered, intentionally laid out. The circular arrangement is thought to have created a cohesive unit for the greatest number of people to face each other.

Most of the rock resources at Bull Brook came from no closer than one hundred and fifty miles away, sourced in New York, Pennsylvania, and New Hampshire. One of the major sources, Musungun chert from northern Maine, came from two hundred and fifty miles away. Each discrete encampment at Bull Brook where these rocks were found was elliptical in shape, about twenty feet across, and contained fire hearths and burned bones. The entire site, at least what was found in a hasty archaeological survey, was about five hundred feet across.

The city we were heading toward in the Black Rock Desert was two miles across and was based on more or less the same footprint as Bull Brook, a ring of camps facing into an open center, each camp with its own purpose and show of artifacts. Like Bull Brook, this city was a single-use location. The next year it would occupy a different part of the playa.

The Bull Brook site is rivaled by a larger aggregation on the High Plains of the Colorado-Wyoming border. This is a fluted stronghold called the Lindenmeier site, the largest known concentration of the Folsom tradition. Folsom was a mobile bison-hunting group that held the middle of the continent around eleven thousand years ago. They were establishing themselves up high against the face of the Rockies, and their fires must have looked like a beacon. The layout and extent of Lindenmeier is unknown, some of it eroded, some buried under ten to twenty feet of sediment. Like Bull Brook, people came from all directions, stone sources arriving from a hundred or more miles away. Kremmling and tiger cherts poured in from a route across the Rockies laid down in the Clovis age. Red ochre was imported from a hundred miles north into Wyoming at the Hell Gap site, an Ice Age ochre quarry that fed many archaeological sites on the High Plains.

Before Lindenmeier and Bull Brook, the late Pleistocene has little evidence of people gathering in extended groups. Mockingbird Gap in the Jornada of southern New Mexico is unusual. Buttermilk Creek in Texas, where flakes, weapons, and tools were stacked on each other for thousands of years back to before Clovis, is even rarer. This appearance of camps gathered together is new. Greg Nunn would say they'd been doing this all along, but this is the first time they rise to the archaeological surface. Monster Slayers had done the hard work, the great and terrifying beasts put down as people teamed up for the next age, the human age.

Archaeologist Jason LaBelle oversees scientific research at the Lindenmeier site in Colorado near the Wyoming border. On a crisp autumn morning, he brought his students out from Colorado State University, showing them the site they'd been studying in the classroom all semester. When he handed out pin flags, the students bobbed like sewing needles, marking flakes of rock and pieces of burnt bone winnowed to the surface. In places the ground looked showered by cultural debris. When he sat on the ground and talked with his students at lunch, LaBelle said, "It must have been some kind of Rainbow Gathering or Burning Man." It was a greater concentration of Ice Age artifacts than you'll find almost anywhere in the Americas.

"Imagine you have campfires up here," LaBelle said, gesturing at the view. "It's a site to see out from, but it's also a sight to be seen. If you wanted to draw people to you, this would be your place."

Lindenmeier was found atop a sort of throne on the land backing up into the Rockies and looking out over the plains, what LaBelle calls a "viewshed." From here, you can see well into Wyoming and down to Pikes Peak standing over Colorado Springs, one hundred and forty miles to the south. It is one of the broadest habitable views in North America.

LaBelle grew up on the flats east of Lindenmeier. He said you could see the site from town, it was "part of people's modern geography." It is the highest point attached to the High Plains, a place where Clovis was based before Folsom. He said that, as a kid, instead of having posters of *Star Wars* and Farrah Fawcett on his bedroom walls, he put up a map of the Lindenmeier site. He knows why it was here: because it could be seen. The rule of ancient people is visibility.

Sitting in the dry autumn grass, he swiped his hand through the air, defining the larger geography. "We are now in the exact center of Folsom territory," he said, cutting the horizon into quadrants. "Canada to Mexico, and Illinois to the west side of Wyoming, we're in the middle of it."

The locations he described are also the ranges of toolstones that ended up here. People were coming from all directions, sewing themselves more permanently into the fabric of time and earth by picking out a place they returned to again and again, leaving abundant archaeology.

LaBelle mentioned obsidian found here, sourced back to Yellowstone, almost four hundred miles to the north, and obsidian from a volcanic caldera in northern New Mexico, almost four hundred miles to the south. Cherts were coming down from the Dakotas, and Edwards Plateau chert up from Texas.

Looking across our viewshed, LaBelle said, "The Great Plains are impenetrable in a way; how do you navigate? Being up here is about navigation, and meeting, and companionship. Even if you've never been here before, you can see the way."

After lunch, LaBelle and his students walked through colored bits and shards of chipped rocks: oranges, yellows, blacks. They found a knocked-out channel flake speaking to the manufacture of small fluted weapons.

A student who'd been working summers as a budding archaeologist at the Hell Gap ochre excavations in Wyoming came upon what looked like eroded bone poking out of the dirt in an arroyo's exposed face. A few more students gathered around the bone, which appeared to be in an exposed dish of ash-colored soil, sure sign of a fire pit. Soon everyone was there, LaBelle advising them to pick just a little at this exposed cross section, grains of soil letting free with tweezers and a pencil-thin bamboo probe meant for this kind of small, exploratory work. It felt like a surgical operation, voices low, assistants stepping in to take artifacts handed to them. Within minutes, they'd uncovered edges of chert flakes, what looked like a piece of charcoal, and more bone as soft as chalk. With only a half-inch of arroyo plucked away, they were opening a window into the Folsom age, a camp buried by eleven thousand years of dust storms and floods, arroyos filling each other with mud and later bisecting themselves, revealing what was buried. The more they found, the slower they picked, talking about a bison hunting camp, no more *latifrons*, now only *antiquus*, a highly adaptable bovid that met the demands of a warmer climate and dangerous atlatl throwers.

As excitement built around this find in the arroyo, LaBelle said, "That's enough."

One student asked if they shouldn't remove the bones and bag them, at least find out what they are. Another asked if there should be a pin flag left behind to mark their discovery. LaBelle said no, just leave it.

As we walked away, he said to me, "Those things are all over the place."

We took one of the higher summits to get a look around. Three days from the city, we had a vantage of a couple thousand feet looking down on blankness, the broad arms of Lake Lahontan wrapping around islands and peninsulas. The lake would have gone past every horizon we could see from here.

The city was swallowed in its own dust storm, industry stirring up the playa. Forklifts, cranes, and front-end loaders were teeming, preparing for the opening of the gates, too far away to pick out with binoculars. All we could see was a shimmering point, a dome of dust miles across. It was a ritual, a festival, an annual gathering, and from up here in a clean wind, it looked too dusty for comfort. The consensus was that we were going, but being alone in the vast arms of Lahontan had its appeal.

Dr. John said, "It's like Obi-Wan and Luke looking over that city in the desert, what was it, a hive of scum and villainy?"

"Mos Eisley," said Q, referring to the scene in *Star Wars*, one of the legends we have been telling in our time.

"They just had to go," said the doctor. "It was their quest."

Our quest was to get to the city. We had tickets to get in. My eye kept darting back to where we'd come from, the sweet isolation we'd been passing through. We were looking down into the dry chap of what is known as the Quinn River, a northern arm of Lahontan that once reached into Oregon. Thirty miles from the city, the Quinn River flows down to this pan of salts of crunchy evaporites, where it disappears, not enough water to feed a lake. From up here it looked like an empty delta, a place you'd expect to see rusted ships keeled on their sides. The opposite shore twenty miles away from us is the foot of the Jackson Mountains, one of the largest islands on Lahontan. A Columbian mammoth skeleton was found on the shores of the Jackson Mountains. It is the largest complete mammoth skeleton in the

United States, excavated in 1982 during a summer of 120-degree heat. A report from the site reads, "A sudden violent sandstorm one day reduced visibility to zero, but caused no damage except to human eyes and skin."

Among the bones of this mammoth, human tools and pieces of stone weapons were found, with some question as to whether they were associated with the mammoth or they just happened to be in the same place at a later date, and the two deflated into each other over time. The excavators also found the milk tooth of a young sabertooth cat, enough to bring up a tantalizing scene, a carcass fought over by two species, two-legged versus four on the remains of an enormous beast, spears and shouting. This mammoth, which has a single contested radiocarbon date of eleven thousand years old, may have been sighted on the shores of a mountainous island in the middle of Lahontan. Hunters swam across the blue water, weapons bundled on their backs, perhaps pursuing the last of the last.

There has been funded, legitimate talk in scientific circles about genetically reviving the mammoth, bringing it back from the other side of extinction. It is being called reparation for a mistake we made as a species, a practice run at bringing back animals we drove out of existence. Though I'd be amazed to see a live mammoth in my own age, I'd be equally horrified. Before waving the flag of what is being called "de-extinction," understand that this would not be an Ice Age mammoth woken to the twenty-first century as if from a block of ice. This would be a "mammophant." Forty-five mammoth genes have been sequenced out of the thousands more needed, and the remainder would come from an Asian elephant.

The horrified part of me asks what danger lies in uncorking, mixing, and letting loose this genetic material, but after corn with chromosomes turned to pesticides and goats producing spiderwebs in their milk, that question feels like it's come and gone. My more immediate concern is what happens when a large, charismatic pioneer is released into a new world, every horizon a possibility. Think zebra mussels, kudzu vines, but with tusks. If it worked and more than a genetic slime in a petri dish was created—if we came up with a creature that was part elephant and part mammoth—what will we have done? The introduction of new species is a primary driver of modern extinctions. Pushing an already extinct animal into a world having trouble holding on to what it already has is an entirely different kind of acceleration, another ball thrown into the circus ring. How long would this invented species last? Beyond the age of governments and Pleistocene parks?

From this summit, I could point one hand at the mammoth find at the foot of the Jackson Mountains, and with the other point 180 degrees away at Burning Man and its rising dust. Towering art sculptures and temples were being built in the desert, seventy thousand participants expected this year, the same year our global population was pushing seven billion. This would become the third-largest city in Nevada, behind Reno and Las Vegas, and it would be ephemeral, rising on the same lake bottom every year in a matter of weeks. Two ends of time felt as if they were balanced across the fulcrum of this rocky summit, ten thousand years in the past on one side and everything to come after on the other. For a moment, all the disjointed pieces of time felt in equilibrium.

The next morning, two days from the city, we were camped on the lake bottom, on the open face of nothing. A storm rolled in before dawn and by first light, thunderheads sparked with lightning. Fat dots of rain fell as the old lakebed woke and began talking back to itself.

I was out of my sleeping bag, crouched with only the contact of my shoe soles on the ground. The air flashed white. Four seconds later, thunder pounded my eardrums. Q would tell me that he saw blue zaps of electricity on some of the gear. I made as little contact as I could, thinking an umbrella would be helpful in the rain, but not wanting to make myself the tallest point on the playa.

The storm hammered the city, now about seven miles away. At dawn, I counted twelve ground-strikes, direct hits inside the enclave of structures, where wooden effigies and temples pointed at the sky like lightning rods. The storm had found its mark. As if the heavens were making a point, the entire sky seemed to be drawn to the city, lightning and rain pulled from all directions and mashed into the ground by the thumb of God. We've only learned the tool of the lever, the advantage of the atlatl, making ourselves larger than we actually are, as weights of weather and climate are being thrown around, the sky doing as it pleases.

Scientists have come together to name the next geologic epoch, a change we are very much part of, not so different from Clovis-age people watching the Ice Age crumble. Many say the Holocene has ended and we are entering a new age. It is a time of widespread extinctions and unraveling climates, superstorms, and inundated cities. We are rolling out this new age with confetti and cheers, cabinets and scientific committees trying to decide if it began twenty-eight hundred years ago, when the iron smelters of Carthage first left carbon layers in the ice caps, or if the Atomic Age marks the real start, the explosion at the Trinity Site kicking us off the ledge. The name of this period doesn't matter so much; historians may find it recorded in the tiniest threads of writing, amused perhaps at our inventiveness. We call it *Anthropocene*. It means *New Man*, the time of humans. I prefer *Hubriscene*.

I'm a fan of the Holocene, the time we're in, or at least have been since the close of the Ice Age. The Pleistocene was a death trap of teeth, claws, and glaciers, while nobody knows what to make of the upcoming Anthropocene. We seem to be bookended by the fierce and the uncharted.

Am I sentimental to hold on to our familiar age? Who is not in love with these blue skies pillared with clouds, and the many species around us, the way more oxygen comes out of the ocean than methane, the deep breathing of the forests and grasslands? I do not want this party to end, yet another one seems to be starting, pushed onstage with zeal.

I've listened to the findings of climate researchers in Greenland as storms howled around our tents, snow building against us. I've gone out with NASA atmospheric researchers to drop remote sensing stations, and have heard how much the climate is evolving and how fast. Ice caps deflate and the Gulf Stream starts a wobble-dance in the Atlantic, then settles back down.

On this thin skin of air, water, and soil, I've read reams of scientific papers, looked at bubbles in unsheathed ice cores. I asked scientists the same question so many times: *Can it be changed, can we redirect this?* The answer was always *yes*. This is nothing but change. Flap your arms, it could have an effect. But can we come back to where we were? That inquiry is met with different answers, mostly *No, you can never go back*. There is only the world ahead.

As the storm let up, we could hear the arrhythmic pounding of

DJs. The city roared and thrummed into a storm-blasted sunrise, as if the lightning had only increased the energy. They were shouting at the sky for more.

By nightfall, all memory of the rain was gone. A bony, dead coyote lay on the pan where it had dropped months ago. Light of the Milky Way powdered across the sky. We'd made it to within a few miles of the city. Now we could see volleys of flame shooting upward. A continuous wave of sound rolled across the desert. Lasers fired out of what in my binoculars looked like the wingless fuselage of a passenger jet that had been trucked in. This was Bull Brook and Lindenmeier on a twenty-first-century scale, something people living along these late Ice Age shores would have understood, at least in concept: *Come together, make yourselves into one.* We were playing out one of our oldest stories, the myth of the eternal return, gathering to mark time and make a ritual of ourselves. This is how people continue. We bring stories to each other, learning what has been witnessed, what the scouts have seen, what new ideas have bubbled up. We show each other our maps, our tools, mixing genes and languages. This has been happening since the beginning.

People with stemmed weapons from the Ice Age wove through those with fluted technology, and out came the classic Indian arrowhead, a small-stemmed projectile affixed to the end of an arrow shaft. The bases of these points were finely knapped into various kinds of barbs, ears, tangs, or notches, possible attributes of the fluted tradition. As the curtain rose on the Holocene, Native America began with an estimated thousand languages, genes going back to Clovis age if not older. The arrowhead that finally appeared, now part of a bow and arrow technology more suited to deer and rabbit, was what came out of this Holocene aggregation, a sort of agreement.

As we stood in the desert at the edge of the burning city, I thought about not going, not being part of the next great aggregation on this lake. We all thought about it. The horizon was ablaze. Lasers struck us where we stood, illuminating the lakebed. It seemed like too much, a grandstanding civilization up on its tiptoes calling everyone it could reach. Is no animal or habitat safe in our path? I wanted to take my tools, my pen and paper, and turn the other way. I didn't want to be party to this very human mania, the unending urges of a species hell-bent on colonizing anything it touches. But I remain a

Historic Native American arrowhead
following the Ice Age, an amalgam of
late Pleistocene traditions in a new age

student of legends. I can't help poking at the stories that I hear, giving them a little flame, sending up sparks.

Etha, from Bali, the most recent arrival among us, took up her fiddle. She ducked away and waltzed across the playa, her song spilling out. It was a gypsy wail, and as she played I thought it was a song I knew in my deepest heart, picking up a stringed instrument, coming to the edge, and plucking a tune. This is how we have taken hold, moving a melody into impossible places, scratching the bones of a mammoth in a remote valley at the beginning of time.

On another lakeshore, a smaller nearby arm, is the series of rock art panels dated to 14,800 years ago. A carbon-bearing mineral formed every time the lake rose and inundated the rock art, leaving a patina that could be dated. The layers of patina span from 14,800 to about 10,500 years ago, glyphs added consecutively every time the lake dropped. Between 14,800 to 13,100 years ago, the boulders were exposed to air and people worked on them. Lake levels rose and submerged the boulders for a couple thousand years, and exposed them again from 11,300 to 10,500 years ago when people picked up the task as if they'd never stopped. This is continuity, a remembered place over the final five thousand years of the Ice Age.

I traveled to this ancient rock art with my family, one of the last road trips we took together, my lap strewn with maps as my wife drove across empty basins and ranges along a Nevada highway. The rock art was not a marked locality. Finding it took some triangulation, reading between the lines of archaeological reports. We had it narrowed down to a sub-basin within Lahontan, an arm of an arm. A winding dirt road led us to a barbed-wire cattle gate, the sign saying to close it behind us. Out of a clutter of empty juice boxes and camping gear we spilled into the Nevada desert.

A spring came out of the mountainside, its source roughly fenced with barbed wire. A clear stream issued from the ground, watercress waving in the current as the spring flowed out of its luscious, vegetated enclosure down to the cattle-smashed slope below. A squared fence was all that kept the spring from becoming a trampled, muddy crater. It was an emerald on the hill, a point of life and yarrow flowers. As we walked by, we stopped for the bumblebees and birds inhabiting this gushing well, this emergence. In the Ice Age it must have roared.

As kids glided ahead through the dry grass of midsummer, we found signs of occupation. Some of the bulbous tufa formations had eroded open and were hollow inside, their floors buried in woodrat droppings, rough ceilings blackened with smoke. At night, these must have looked like a hive, lights flickering atop each other at the edge of a lake, people living as if in a clutch of grapes. Paisley, in Oregon, would have looked similar, the caves lit along the edge of their own lake at about the same time, give or take a few hundred centuries.

The tufa is like pumice, easy to etch or scratch. We found a flat-faced boulder of this rock that had been deeply crosshatched by human tools. Well worn, the geometric grooves were lightly coated with minerals from the rise and fall of the lake behind it now dried to a salty crisp. Both kids were mildly impressed but more interested in tearing off, playing tag and racing across the slope. Here people had been making the first banners of themselves, the original drive-in theaters. They had stood in this same place, their feet beneath ours. Jado, with a sixteen-stitch Alaskan scar on his forehead, looked at the rock art and said, "I don't know what's the big deal about a bunch of X's," and he ran away.

I felt as if I'd been going from landmark to landmark, asking, *Are you my mother?* Like a lost child on a lost continent, I searched to root myself and find solid ground. I visited the oldest sites as if on a pilgrimage, stopping to recognize those who came before, lineages upon lineages, arrivals upon arrivals leading back to the first flicker of a campfire, a panel of figures etched onto clusters of tufa boulders in Nevada, one of the earliest human lights on the continent.

Another stop I wanted to make was the Black Rock Desert, up an arm of Lake Lahontan an hour or two away from this panel. Our task as a family was filling in maps, following back roads, unfamiliar turns, high on D4 or whatever mythology of childhood we told ourselves. The next day, we hit the Black Rock playa on a road that became no road, a dusty pan with nothing on it, no reason to drive in one place or another, the ground too hard to take a tire track. Construction had yet to begin on the annular city that would rise here. The lake bottom of Lahontan was still empty.

We triangulated on last year's city the same way we had on the rock art, and I asked to be let off. I wanted a few minutes alone on the playa, breathing on this lake bottom.

My wife stopped to let me out. Kids climbed through their windows, out onto the running boards, and banged car metal with their hands when they were ready. The SUV rolled away like an amusement park ride, their mother at the wheel, leaving me standing on the open, white plain. The vehicle became smaller, shouts of glee fading to a pinprick, almost silence. The ground was cracked and featureless. I thought there had to be something, an artifact, a spray of glitter, or charcoal from the giant wooden man they would build and burn to the ground, but I saw nothing. The site had been thoroughly cleaned, the city erased, maddening for future archaeologists.

I looked for the throngs in my imagination, conjuring people, dust, and fire. Half-naked crowds rose from the ground. It was the center of the city and a giant wooden man, effigy of ourselves, burned with ferocity, showering the stars with plumes of sparks. I walked through the apparition of this city, aware of what happened as if we'd been here before, playing out this story like we play out all our stories, coming back and pounding the earth with our feet.

I listened for them, but heard only my steps across playa dust. For all we do, we vanish this easily, opening a space at the end of one age to send our children into the beginning of the next. The dot of the SUV swung around a couple miles out. It became larger and I began to hear the engine and the whoops of two boys coming back to me. They hopped off before a full stop, testing their bravery as their feet hit the ground running. Given the extra momentum, they sprinted like fawns across the expanse, leaping and shouting, the sounds of their joy fading into desert air.

EPILOGUE

The picture that came across Alexander Rose's cell phone looked like a blackness within blackness, a hole within a hole. He said that excavators in West Texas had been blasting a tunnel into solid limestone when they came across a cavern never before opened to outside air, a secret space within the earth. They'd taken a picture of their find and sent it to him.

Rose said this new void could hold up progress, or at least make the project more interesting. A giant clock was being built and it would be placed inside the chamber they were excavating, where it would tick-tock gently for the next ten thousand years. He said they didn't know how deep this new cave went, but workers told him when they shouted into it, it made their voices echo.

I think of Raven, of course, and the first crack of light that entered the clamshell at the beginning of time. Raven opened the darkness and out we poured.

Rose was the project manager for this mechanism being built of titanium, ceramic, and stone, materials chosen so that they would not rust or fuse together over time. He said it was not a time capsule per se. It was a clock, a message to the future and a way for us now to understand the larger time frames we're involved with, responsibilities that last beyond lifetimes. It would run off pressure bellows rising and falling with the sun, turning an array of gears, some large enough they could be running Big Ben. Its counterweights are five tons each and are the size of cars.

By ten thousand years ago, climates had turned toward a warm and rainy start of the Holocene. The Americas were fully occupied, nearly every decent rockshelter bearing toolstone and charcoal. Mammoths, sabertooth cats, and dire wolves had been gone for long enough they would have been myth, turned into tales of monsters that once ruled the world. Eating smaller game in the Holocene, in some places eating more grass seed and root than meat, and painting ghostly imagery on canyon walls with red ochre, these people came from the first people, the original ancestry, birth of Native America.

Rose and I were in a Northern California warehouse where clock parts were being built in an industrial district north of the Golden Gate Bridge. Mechanisms were separated, gears in clusters suspended from rafters where engineers had been installing, then carefully aligning the components. The workspace hummed from stress tests on ball bearings sped up to match ten thousand years of rotations, seeing if they could withstand the wear. Designers were serious about the machine working that long. Once underground, the clock would have an entrance and exit for visitors to pass through. He said the idea isn't so much to have visitors anytime soon, but in the future—might be centuries from now, might be thousands of years.

Besides this new cave problem, he was working with artists and linguists, trying to come up with what to put on signs that would explain how the clock functions. The idea is that people would enter it, read the instructions, and calibrate the dates properly, causing the clock to chime like a giant music box.

Rose didn't think we'd be speaking English ten thousand years from now. In several hundred years, the word on the page would require translation. Definitions and spellings would change, new words brought in, altering the way sentences are spoken and ideas conveyed. The way we conceive time and space, the flow of events, is conceded into what we say, how we say it. The clock will reach beyond that. He waved his phone in the air, indicating how easy it is to become trapped in the small, digital box of *now* thinking this is all there is.

He told me that a spiral staircase would rise through the interior of this giant clock. The staircase would be unlit, in near complete darkness. Rose sees people coming as a pilgrimage, torches or headlamps or whatever form of illumination might be used for winding into the well of the machine.

Rose's phone rang again, a call from the excavators. He said he had to take it, they'd be wondering if work should resume. As he walked away, I ran my fingers along the cool titanium teeth of a gear. Rose had welcomed my touching them, letting me give them a nudge.

Several inches of turn had budged the whole thing. So incredibly balanced was it all that as soon as I moved one, other wheels began moving across the contraption, action and instant reaction.

The largest assembly of gears hanging from the rafters was based on a star-shaped center known as a *Geneva drive*, a technology used originally in Swiss watches. This is a circle and square combined, a device that transfers continuous motion into incremental strokes. It is the effortless cruise of the sun broken down to the tick marks of seconds. You might say it's how we invented time.

The ten-thousand-year clock would be our artifact left in a cave. To understand who was here, you might dig it out and study its pieces. Another site in the Nevada desert was already being selected for a hypothetical second clock, as if we were seeding the land, adding our version of Clovis caches under boulders. It may be a message from an ineffable past for people in the future, their ear pressed into a can attached to a string where they listen for our lips to move.

I would tell them of the roaring overpasses around this warehouse, their stark shadows cast onto streets below, abandoned shopping carts, plastic drink cups smashed on asphalt, cars parked for so long their undercarriages are laced with spiderwebs. This is our atlas, a world that will be lost by the time this clock winds down. What kind of lost, I can't say. Whoever returns to this place might think it was a blink of the eye, or they might marvel at the ingenuity of the makers, wondering what ancient people had come this way.

I would tell them what was indelible in our time, the deeply forested hills above the city, sea crashing to mist on a rocky coast, rivers dancing through granite of the Sierra. These will remain, I pray.

I reached in, grasped the biggest gear, and gave it a good, hard spin. Cogs bit into each other. Wheels turned. The machine woke. As smooth as breathing, time began to tick.

Walking. I am listening to a deeper way.
Suddenly all my ancestors are behind me.

—LINDA HOGAN

ACKNOWLEDGMENTS

Many were leaned on to make this book happen, kindnesses too numerous to count. Thank you to all who were patient, instructive, or one way or another helped the stories flow.

I am indebted to those who directly made it possible for words to land on these pages. My editor, Dan Frank, kept the faith while I was lost in the Ice Age, no telling when I'd surface. I so much appreciate his calmness and wise direction, sitting beside me on a Central Park bench to hear my latest escapade and urging me on. Meanwhile, assistant editor Betsy Sallee made the complexities of publication as easy as possible for a bear of very little brain. Thank you to copy editor Diane Sylvain, who took a small screwdriver and tightened the work just so, and production editor Rita Madrigal, who carried the platter of this book out the swinging doors while I kept adding dashes and garnish.

Working with the illustrator, Sarah Gilman, was a joy, talking out possibilities and gleefully awaiting each of her drawings to ping my inbox just as she finished them.

To Kathy Anderson, who originally drew the idea of this book out of me, and for our time together creating and un-creating, thank you.

Thanks to my close readers Jerry Roberts, Felix Marti, and Tim Goncharoff for throwing pages and scribbles right back at me. The tag team of writer Amy Irvine and poet Daiva Chesonis, sharing space at the kitchen table, reading our stories back and forth to one another, lifted my soul. For my children, Jasper and Jado, around whom this book is shaped, who started the fire when I was too busy, and who looked over my shoulder offering their own surprisingly useful edits, I am forever grateful.

Finally, my thanks go to Susie Billings and Duncan Ferguson for the hours they both spent curled up listening as I read aloud. I will cherish every candlelit evening while their sweet dog, Axel, listened in, commenting on the manuscript in his own canid way.

NOTES

2. INNER BERINGIA

22 "Nowadays we do not see": Osada and Endo, "Unicorn, Mammoth, Whale," 199. (Originally recorded by Hungarian folklorist Bernát Munkácsi, this story was later adapted by Mansi linguist Béla Kálmán in his 1963 book *Chrestomathia Vogulica*.)

23 "just one of many": Hart and Sussman, *Man the Hunted*, xvi.

25 "a colossal omnivorous bear": Figueirido and Pérez-Claros, et al., "Demythologizing *Arctodus simus*," 272.

26 "wolves, lions": Bourgeon, "Earliest Human Presence in North America Dated to the Last Glacial Maximum," 6.

29 "eastern Beringia offered": Cinq-Mars and Morlan, "Bluefish Caves and Old Crow Basin," 9.

31 "The archaeozoological evidence": Bourgeon, "Earliest Human Presence in North America Dated to the Last Glacial Maximum," 6.

3. HOUSE OF ICE

53 "Brains exist because": Allman, *Evolving Brains*, 2.

53 "extend the operation of memory": Vygotsky, *Mind in Society*, 39.

59 "Prehistory has produced": Wreschner, Bolton, Butzer, et al., "Red Ochre and Human Evolution," 631.

4. THE LONG COAST

67 "These would have been": Fladmark, "Alternate Migration Corridors for Early Man in North America," 64.

67 "By about 16,000 years ago": Erlandson, Graham, Bourque, et al., "The Kelp Highway Hypothesis," 162.

70 "Equating women solely": Waguespack, "The Organization of Male and Female Labor in Foraging Societies," 674.

94 "Assuming that other late Pleistocene": Dillehay, Ramírez, Pino, et al., "Monte Verde," 786.

94 "It is quite possible that the Americas": Surovell, "Early Paleoindian Women, Children, Mobility, and Fertility," 493.

94 "achieved long-term rates": Bettinger, "Prehistoric Hunter-Gatherer Population Growth Rates Rival Those of Agriculturalists," 812.

98 "a steppe-like grassland environment": Steffen and Harington, "Giant Short-Faced Bear (*Arctodus simus*) from Late Wisconsinan Deposits at Cowichan Head, Vancouver Island, British Columbia," 1034.

5. PLAYGROUND OF GIANTS

101 "a Mr. Stanley": Jefferson, *Notes on the State of Virginia*, 43.

102 "very large kind of animal": Teit, "Kaska Tales," 446.

102 "was searching for a wife": Strong, "North American Indian Traditions Suggesting a Knowledge of the Mammoth," 85.

106 "Impact points on mammoth": Holen and Holen, "Evidence for a Human Occupation of the North American Great Plains During the Last Glacial Maximum," 85.

108 "utilized carcasses more fully": Valkenburgh and Hertel, "Tough Times at La Brea," 456.

7. A DANGEROUS EDEN

131 "the drawing was made prior": Purdy, Jones, et al., "Earliest Art in the Americas," 2913.

134 "man's best friend": Fiedel, "Man's Best Friend—Mammoth's Worst Enemy?," 11.

134 "harried native carnivores": Ibid., 17.

136 "The mammoth lives": Osada and Endo, "Unicorn, Mammoth, Whale," 199.

141 "the well-known phenomenon": Straus, Meltzer, and Goebel, "Ice Age Atlantis?," 507.

153 "the erroneous impression": Frison, *Survival by Hunting*, 34.

154 "I strongly believe": Ibid., 41.

162 "In the low diversity grasslands": Hill, "A Moveable Feast," 417.

163 "especially prolonged desiccation": Faught, "The Underwater Archaeology of Paleolandscapes, Apalachee Bay, Florida," 283.

8. CULT OF THE FLUTED POINT

170 "At the quarry": Speth, Newlander, White, et al., "Early Paleoindian Big-Game Hunting in North America," 115.

9. THE LAST MAMMOTH HUNT

180 "the highest frequency": Surovell and Waguespack, "How Many Elephant Kills Are 14?," 83.

182 "When the societal stress": Bradley and Collins, "Imagining Clovis as a Cultural Revitalization Movement," 252.

184 "represent either discrete campsites": Graham, "Mockingbird Gap," 8.

185 "initial colonization of the intermountain region": Beck and Jones, "Clovis and Western Stemmed," 81.

187 "extinct elephant bones": Boldurian, "Memorial," 86.

187 "Indian warheads": Ibid.

189 "My Yup'ik name is *Agragiiq*": Hovey and Trop, "Gambell Teenager Leads Successful Whale Hunt," 1.

190 "While the vast majority": Haynes, *American Megafaunal Extinctions at the End of the Pleistocene*, 79.

10. AMERICAN BABYLON

209 "high-quality, exotic toolstones": Lambert and Loebel, "Paleoindian Colonization of the Recently Deglaciated Great Lakes," 285.

11. THE PARTY AT THE BEGINNING OF THE WORLD

215 "later Holocene crania": Faught, "Archaeological Roots of Human Diversity in the New World," 687–8.

226 "A sudden violent sandstorm": Stout, "Discovery and C14 Dating of the Black Rock Desert Mammoth," 22.

SELECTED BIBLIOGRAPHY

Adams, K.D., Goebel, T., Graf, K., et al. 2008. "Late Pleistocene and Early Holocene Lake-Level Fluctuations in the Lahontan Basin, Nevada: Implications for the Distribution of Archaeological Sites." *Geoarchaeology* 23(5): 608–643.

Ager, T.A. 2003. "Late Quaternary Vegetation and Climate History of the Central Bering Land bridge from St. Michael Island, Western Alaska." *Quaternary Research* 60: 19–32.

Ahler, S.A. 2000. "Why Flute? Folsom Point Design and Adaptation." *Journal of Archaeological Science* 27: 799–820.

Alexon, R. and Dunbar, J.S. 1984. "A *Bison antiquus* Kill Site, Wacissa River, Jefferson County, Florida." *American Antiquity* 49: 384–392.

Allman, J. 1999. *Evolving Brains.* New York: Scientific American Library.

Ambler, B.M. 1999. "Folsom Chipped Stone Artifacts from the Lindenmeier Site, Colorado: The Coffin Collection." M.A. thesis, Department of Anthropology, Colorado State University, Fort Collins.

Anderson, D.G., Goodyear, A.C., Kennett, J., et al. 2011. "Multiple Lines of Evidence for Possible Human Population Decline/Settlement Reorganization During the Early Younger Dryas." *Quaternary International* 242: 570–583.

Arroyo-Cabralesa, J., Polaco, O.J., et al. 2006. "A Preliminary View of the Coexistence of Mammoth and Early Peoples in México." *Quaternary International* 142–143: 79–86.

Ballenger, J.A.M., Holliday, V.T., Kowler, A.L., et al. 2011. "Evidence for Younger Dryas Global Climate Oscillation and Human Response in the American Southwest." *Quaternary International* 242: 502–519.

Bamforth, D.B. 2011. "Origin Stories, Archaeological Evidence, and Postclovis Paleoindian Bison Hunting on the Great Plains." *American Antiquity* 76(1): 24–40.

———.2009. "Projectile Points, People, and Plains Paleoindian Perambulations." *Journal of Anthropological Archaeology,*28: 142–157.

Basilyan, A.E., Anisimov, M.A., Nikolskiy, P.A., et al. 2011. "Wooly Mammoth Mass Accumulation Next to the Paleolithic Yana RHS Site, Arctic Siberia: Its Geology, Age, and Relation to Past Human Activity." *Journal of Archaeological Science* 38: 2461–2474.

Beck, C., Jones, G.T. 2010. "Clovis and Western Stemmed: Population Migration and the Meeting of Two Technologies in the Intermountain West." *Society for American Archaeology* 75(1): 81–116.

Beck, C., Taylor, A.K., Jones, G.T., et al. 2002. "Rocks Are Heavy: Transport Costs and Paleoarchaic Quarry Behavior in the Great Basin." *Journal of Anthropological Archaeology* 21: 481–507.

Berger, J., Swenson, J.E., Persson, I. 2001. "Recolonizing Carnivores and Naïve Prey: Conservation Lessons from Pleistocene." *Science, New Series* 291(5506): 1036–1039.

Bettinger, R.L. 2016. "Prehistoric Hunter-Gatherer Population Growth Rates Rival Those of Agriculturalists." *Proceedings of the National Academy of Sciences* 113(4): 812–814.

Boldurian, A.T. 2004. "Memorial: James Ridgley Whiteman 1910–2003." *Plains Anthropologist* 49(189): 85–90.

Bourgeon, L. 2015. "Bluefish Cave II (Yukon Territory, Canada): Taphonomic Study of a Bone Assemblage." *PaleoAmerica* 1(1): 105–108.

Bourgeon, L., Burke, A., et al. 2017. "Earliest Human Presence in North America Dated to the Last Glacial Maximum: New Radiocarbon Dates from Bluefish Caves, Canada." *PLoS One* 12(1): e0169486; doi: 10.1371/journal.pone.0169486.

Bradley, B. and Collins, M. 2014. "Imagining Clovis as a Cultural Revitalization Movement." In *Paleoamerican Odyssey*, K.E. Graf, C.V. Ketron, and M.R. Waters, eds. College Station: Texas A&M University Press, 247–256.

Briles, C.E., Whitlock, C., Meltzer, D.J. 2012. "Last Glacial–interglacial Environments in the Southern Rocky Mountains, USA and Implications for Younger Dryas-age Human Occupation." *Quaternary Research* 77: 96–103.

Buchanan, B., Kilby, J.D., Huckell B.B., et al. 2012. "A Morphometric assessment of the Intended Function of Cached Clovis Points." PLoS ONE 7(2): e30530. doi:10.1371/journal.pone.0030530.

Buchanan, B., O'Brien, M.J., Collard, M. 2014. "Continent-wide or Region-specific? A Geometric Morphometrics-based Assessment of Variation in Clovis Point Shape." *Archaeological and Anthropological Sciences* 6:145–162.

Buehler, D.M., Baker, A.J., and Piersma, T. 2006. "Reconstructing Palaeoflyways of the Late Pleistocene and Early Holocene Red Knot *Calidris canutus*." *Ardea* 94(3): 485–498.

Byers, D.A., Ugan, A. 2005. "Should We Expect Large Game Specialization in the Late Pleistocene? An Optimal Foraging Perspective on Early Paleoindian Prey Choice." *Journal of Archaeological Science* 32: 1624–1640.

Camp, A.J. 2009. "Pre-archaic Occupations in the West Arm of the Black Rock Desert." M.A. thesis, University of Nevada, Reno.

Cannon, M.D., Meltzer, D.J. 2004. "Early Paleoindian Foraging: Examining the Faunal Evidence for Large Mammal Specialization and Regional Variability in Prey Choice." *Quaternary Science Reviews* 23: 1955–1987.

Carrara, P.E., et al. 2007. "Possible Refugia in the Alexander Archipelago of Southeastern Alaska During the Late Wisconsin Glaciation." *Canadian Journal of Earth Sciences* 44: 229–244.

Cassidy, J., Raab, L.M. 2004. "Boats, Bones, and Biface Bias: The Early Holocene Mariners of Eel Point, San Clemente Island, California." *Society for American Archaeology* 69 (1): 109–130.

Chen, C.S., Burton, M., Greenberger, E., et al. 1999. "Population Migration and the Variation of Dopamine D4 Receptor (DRD4) Allele Frequencies Around the Globe." *Evolution and Human Behavior* 20: 309–324.

Cinq-Mars, J. 1979. "Bluefish Cave I: A Late Pleistocene Eastern Beringian Cave Deposit in the Northern Yukon." *Canadian Journal of Archaeology* 3: 1–32.

Cinq-Mars, J. and Morlan, R.E. 1999. "Bluefish Caves and Old Crow Basin: A New Rapport." In *Ice Age Peoples of North America: Environments, Origins, and Adaptations of the First Americans*, Bonnichsen, R. and Turnmire, K.L., eds. Corvallis: Oregon State University Press, 200–212.

Clark, J., Mitrovica, J.X., et al. 2014. "Coastal Paleogeography of the California–Oregon–Washington and Bering Sea Continental Shelves During the Latest Pleistocene and Holocene: Implications for the Archaeological Record." *Journal of Archaeological Science* 52: 12–23.

Deller, B.D., Ellis, C.J., Keron, J.R. 2009. "Understanding Cache Variability: A Deliberately Burned Early Paleoindian Tool Assemblage from the Crowfield Site, Southwestern Ontario, Canada." *American Antiquity* 74(2): 371–397.

Dillehay, T.D., Ocampo, C., Saavedra, J., et al. 2015. "New Archaeological Evidence for an Early Human Presence at Monte Verde, Chile." *PLoS ONE* 10(12): e0145471. doi.org/10.1371/journal.pone.0141923.

Dillehay, T.D., Ramírez, C., Pino, M., Collins, M. B., et al. 2008. "Monte Verde: Seaweed, Food, Medicine, and the Peopling of South America." *Science* 320: 784–786.

Dixon, J.E. 2013. "Late Pleistocene Colonization of North America from Northeast Asia: New Insights from Large-scale Paleogeographic Reconstructions." *Quaternary International* 285: 57–67.

Donohue, S.L., DeSantis, L.R.G., Schubert B.W., et al. 2013. "Was the Giant Short-Faced Bear a Hyper-Scavenger? A New Approach to the Dietary Study of Ursids Using Dental Microwear Textures." *PLoS ONE* 8(10): e77531. doi:10.1371/journal.pone.0077531.

Dyke, A.S., et al. 2002. "The Laurentide and Innuitian Ice Sheets During the Last Glacial Maximum." *Quaternary Science Review* 21: 9–31.

Elias, S.A., Berman, D., and Alfimov, A. 2000. "Late Pleistocene Beetle Faunas of Beringia: Where East Met West." *Journal of Biogeography* 27: 1349–1363.

Ellis, C.J., Carr, D.H., Loebel, T.J. 2011. "The Younger Dryas and Late Pleistocene Peoples of the Great Lakes Region." *Quaternary International* 242: 534–545.

Erlandson, J.M., Braje, T.J. 2011. "From Asia to the Americas by Boat? Paleogeography, Paleoecology, and Stemmed Points of the Northwest Pacific." *Quaternary International* 239: 28–37.

Erlandson, J.M., Graham, M.H., Bourque, B.J., et al. 2007. "The Kelp Highway Hypothesis: Marine Ecology, the Coastal Migration Theory, and the Peopling of the Americas." *The Journal of Island and Coastal Archaeology* 2(2): 161–174.

Erlandson, J.M., Rick, T.C., Braje, T.J., et al. 2011. "Paleoindian Seafaring, Maritime Technologies, and Coastal Foraging on California's Channel Islands." *Science* 331: 1181–1185.

Eshleman, J.A., Malhi, R.S., Johnson, J.R., et al. 2004. "Mitochondrial DNA and Prehistoric Settlements: Native Migrations on the Western Edge of North America." *Human Biology* 76(1): 55–75.

Faith, J.T. 2011. "Late Pleistocene Climate Change, Nutrient Cycling, and the Megafaunal Extinctions in North America." *Quaternary Science Reviews* 30: 1675–1680.

Faught, M.K. 2008. "Archaeological Roots of Human Diversity in the New World: A Compilation of Accurate and Precise Radiocarbon Ages from Earliest Sites." *American Antiquity* 73(4): 670–698.

———. 2004. "The Underwater Archaeology of Paleolandscapes, Apalachee Bay, Florida." *American Antiquity* 69(2): 275–289.

Feathers, J.K., Rhodes, E.J., Huot, S., et al. 2006. "Luminescence Dating of Sand Deposits Related to Late Pleistocene Human Occupation at the Cactus Hill Site, Virginia, USA." *Quaternary Geochronology* 1(3): 167–187.

Fiedel, S.J. 2005. "Man's Best Friend—Mammoth's Worst Enemy? A Speculative Essay on the Role of Dogs in Paleoindian Colonization and Megafaunal Extinction." *World Archaeology* 37(1): 11–25.

Figueirido, B., Pérez-Claros, J.A., et al. 2010. "Demythologizing *Arctodus simus*, the 'Short-faced' Long-legged and Predaceous Bear That Never Was." *Journal of Vertebrate Paleontology* 30(1): 262–275.

Fisher, D.C. 1984. "Taphonomic Analysis of Late Pleistocene Mastodon Occurrences: Evidence of Butchery by North American Paleo-Indians." *Paleobiology* 10(3): 338–357.

Fitzhugh, B., Shubin, V.O., et al. 2002. "Archaeology in the Kuril Islands: Advances in the Study of Human Paleobiogeography and Northwest Pacific Prehistory." *Arctic Anthropology* 39(1–2): 69–94.

Fix, A.G., 2005. "Rapid Deployment of the Five Founding Amerind mtDNA Haplogroups Via Coastal and Riverine Colonization." *American Journal of Physical Anthropology* 128: 430–436.

Fladmark, K.R., 1979. "Routes: Alternate Migration Corridors for Early Man in North America." *American Antiquity* 44(1): 55–69.

Frison, G.C. 2004. *Survival by Hunting: Prehistoric Human Predators and Animal Prey.* Berkeley: University of California Press.

———. 1998. "Paleoindian Large Mammal Hunters on the Plains of North America." *Proceedings of the National Academy of Sciences* 95: 14576–14583.

Gilligan, I. 2010. "The Prehistoric Development of Clothing: Archaeological Implications of a Thermal Model." *Journal of Archaeological Method and Theory* 17: 15–80.

Goebel, T., Hockett, B., Adams, K.D., et al. 2011. "Climate, Environment, and Humans in North America's Great Basin During the Younger Dryas, 12,900–11,600 Calendar Years Ago." *Quaternary International* 242: 479–501.

Goebel, T., Waters, M.R., Dikova, M., 2003. "The Archaeology of Ushki Lake, Kamchatka, and the Pleistocene Peopling of the Americas." *Science* 301: 501–505.

Graham, D. 2008. "Mockingbird Gap: A Mid-Century Discovery Gets Another Spin." *Mammoth Trumpet* 23(4): 5–16.

Haile, J., Froese, D.G., MacPhee, R.D.E, et al. 2009. "Ancient DNA Reveals Late Survival of Mammoth and Horse in Interior Alaska." *Proceedings of the National Academy of Sciences* 106(2): 22352–22357.

Hamilton, M.J., Buchanan, B. 2010. "Archaeological Support for the Three-Stage Expansion of Modern Humans Across Northeastern Eurasia and into the Americas." *PLoS ONE* 5(8): e12472. doi:10.1371/journal.pone.0012472.

Hare, B., Wobber, V., Wrangham, R. 2012. "The Self-domestication Hypothesis: Evolution of Bonobo Psychology Is Due to Selection Against Aggression." *Animal Behavior* 83(3): 573–585.

Harington, C.R. 2011. "Pleistocene Vertebrates of the Yukon Territory." *Quaternary Science Reviews* 30: 2341–2354.

Hart, D. and Sussman, R.W. 2005. *Man the Hunted: Primates, Predators, and Human Evolution.* New York: Westview.

Haynes, G. 2009. *American Megafaunal Extinctions at the End of the Pleistocene.* Dordrecht: Springer.

———. 2002. "Archaeological Methods for Reconstructing Human

Predation on Terrestrial Vertebrates." *The Paleontological Society Papers* 8: 51–68.

Heaton, T.H., Talbot, S.L., and Shields, G.F. 1996. "An Ice Age Refugium for Large Mammals in the Alexander Archipelago, Southeastern Alaska." *Quaternary Research* 46(2): 186–192.

Hey, J. 2005. "On the Number of New World Founders: A Population Genetic Portrait of the Peopling of the Americas." *PLoS Biology* 3(6): e193. doi:10.1371/journal.pbio.0030193.

Hill Jr., M.E. 2008. "Variation in Paleoindian Fauna Use on the Great Plains and Rocky Mountains of North America." *Quaternary International* 191: 34–52.

———. 2007. "A Moveable Feast: Variation in Faunal Resource Use among Central and Western North American Paleoindian Sites." *Society for American Archaeology* 72(3): 417–438.

Holen, S.R. 2006. "Taphonomy of Two Last Glacial Maximum Mammoth Sites in the Central Great Plains of North America: A Preliminary Report on La Sena and Lovewell." *Quaternary International* 142–143: 30–43.

Holen, S.R., Deméré, T.A., Fisher, D.C., et al. 2016. "A 130,000-year-old Archaeological Site in Southern California, USA." *Nature* 544: 479–483.

Holen, S.R. and Holen, K.A. 2009. "Evidence for a Human Occupation of the North American Great Plains During the Last Glacial Maximum." In: *IV Simposio Internacional Hombre Temprano en América*, eds. Jiménez, J., Polaco, O., Martínez, G., et al. Mexico City and Saltillo: Instituto Nacional de Antropología e Historia, Instituto de Investigaciones Antropológicas de la Universidad Nacional Autónoma de México, Museo del Desierto, 85–106.

Holliday, V.T., Huckell, B.B., Weber, R.H., et al. 2009. "Geoarchaeology of the Mockingbird Gap (Clovis) Site, Jornada del Muerto, New Mexico." *Geoarchaeology* 24(3): 348–370.

Hovey, D. and Trop, K. April 21, 2017. "Gambell Teenager Leads Successful Whale Hunt, Brings Home 57-Foot Bowhead." *KNOM Radio Mission*, Nome, AK.

Huckell, B.B., Holliday, V.T., Weber, R.H. 2007. "Test Investigations at the Mockingbird Gap Clovis Site: Results of the 2006 Field Season." *Current Research in the Pleistocene* 24: 102–104.

Jefferson, T. 1853. *Notes on the State of Virginia*. Richmond: J.W. Randolph.

Jenkins, D.L., Davis, L.G., Stafford T.W., et al. 2014. "Geochronology, Archaeological Context, and DNA at the Paisley Caves." In *Paleoamerican Odyssey*, K.E. Graf, C.V. Ketron, and M.R. Waters, eds. College Station: Texas A&M University Press, 485–510.

Jones, K.L., Krapu, G.L., Brandt, D.A. 2005. "Population Genetic Structure in Migratory Sandhill Cranes and the Role of Pleistocene Glaciations." *Molecular Ecology* 14(9): 2645–2657.

Kauman, D.S., Manley, W.F. 2004. "Pleistocene Maximum and Late Wisconsinan Glacier Extents Across Alaska, U.S.A." *Developments in Quaternary Science* 2: 9–27.

Kivisild, T.E., Reidla, T., Metspalu, M., et al. 2007. "Beringian Standstill and Spread of Native American Founders." *PLoS ONE* 2(9): 10.1371/journal.pone.0000829.

Kooyman, B., Hills, L.V., McNeil, P., et al. "Late Pleistocene Horse Hunting at the Wally's Beach Site (DhPg-8), Canada." *Society for American Archaeology* 71(1): 101–121.

Kuzmin, Y.V. and Keates, S.G. 2005. "Dates Are Not Just Data: Paleolithic Settlement Patterns in Siberia Derived from Radiocarbon Records." *Society for American Archaeology* 70(4): 773–789.

Lahaye, C., Hernandez, M., Boëda E., et al. 2013. "Human Occupation in South America by 20,000 BC: The Toca da Tira Peia Site, Piauí, Brazil." *Journal of Archaeological Science* 40: 2840–2847.

Lambert, J.M. and Loebel, T.J. 2015. "Paleoindian Colonization of the Recently Deglaciated Great Lakes: Mobility and Technological Organization in Eastern Wisconsin." *PaleoAmerica* 1(3): 284–288.

Lee, O.Y.-C, Wu H.H.T., Donoghue, H.D., et al. 2012. "Mycobacterium Tuberculosis Complex Lipid Virulence Factors Preserved in the 17,000-year-old Skeleton of an Extinct Bison, *Bison antiquus*." *PLoS ONE* 7(7): e41923. doi:10.1371/journal.pone.0041923.

Loehle, C. 2007. "Predicting Pleistocene Climate from Vegetation in North America." *Climates of the Past* 3: 109–118.

Lozhkin, A.V., Anderson, P., Eisner, W., et al. 2011. "Late Glacial and Holocene Landscapes of Central Beringia." *Quaternary Research* 76: 383–392.

Lucas, S.G., Morgan, G.S., Hawley, J.W., et al. 2002. "Mammal Footprints from the Upper Pleistocene of the Tularosa Basin, Doña Ana County, New Mexico." In *Geology of White Sands, New Mexico Geological Society Guidebook, 53rd Field Conference,*: 285–288.

Mandryk, C.A.S., Josenhans, H., Fedje, D.W., et al. 2001. "Late Quaternary Paleoenvironments of Northwestern North America: Implications for Inland Versus Coastal Migration Routes." *Quaternary Science Reviews* 20: 301–314.

Matheus, P.E. 1995. "Diet and Co-ecology of Pleistocene Short-faced Bears and Brown Bears in Eastern Beringia." *Quaternary Research* 44(3): 447–453.

McNeil, P., Hills L.V., Kooyman, B., et al. 2005. "Mammoth Tracks Indicate a Declining Late Pleistocene Population in Southwestern Alberta, Canada." *Quaternary Science Reviews* 24: 1253–1259.

Meiri, M., et al. 2014. "Faunal Record Identifies Bering Isthmus Conditions as Constraint to End-Pleistocene Migration to the New World." *Proceedings of the Royal Society B* 281: 1–9.

Meltzer, D.J., Holliday, V.T. 2010. "Would North American Paleoindians Have Noticed Younger Dryas Age Climate Changes?" *Journal of World Prehistory* 23(1): 1–41.

Middleton, E.S., Smith, G.M., Cannon, W.J., et al. 2014. "Paleoindian Rock Art: Establishing the Antiquity of Great Basin Carved Abstract Petroglyphs in the Northern Great Basin." *Journal of Archaeological Science* 43: 21–30.

Miotti, L.L. 2003. "Patagonia: A Paradox for Building Images of the First Americans During the Pleistocene/Holocene Transition." *Quaternary International* 109–110: 147–173.

Misartia, N., Finney, P.B., Jordan, J.W., et al. 2012. "Early Retreat of the Alaska Peninsula Glacier Complex and the Implications for Coastal Migrations of First Americans." *Quaternary Science Reviews*, 48: 1–6.

Morgan, G.S. 2002. "Late Rancholabrean Mammals from Southernmost Florida, and the Neotropical Influence in Florida Pleistocene Faunas." In Emry, R.J. ed., *Cenozoic Mammals of Land and Sea: Tributes to the Career of Clayton E. Ray. Smithsonian Contributions to Paleobiology* (Washington, DC: Smithsonian Institution Press), 15–38.

Morgan, G.S., Lucas, S.G. 2002. "Pleistocene Vertebrates from the White Sands Missile Range, Southern New Mexico." In *Geology of White Sands, New Mexico Geological Society Guidebook, 53rd Field Conference*, 267–276.

Moss, M.L, Erlandson, J.M. 2013. "Waterfowl and Lunate Crescents in Western North America: The Archaeology of the Pacific Flyway." *Journal of World Prehistory* 26: 173–211.

Newby, P., Bradley, J., Spiess, A., et al. 2005. "A Paleoindian Response to Younger Dryas Climate Change." *Quaternary Science Reviews* 24: 141–154.

Osada, T. and Endo, H. 2011. "Unicorn, Mammoth, Whale: Mythological and Etymological Connections of Zoonyms in North and East Asia." *Occasional Paper* 12. Kyoto: Research Institute for Humanity and Nature.

Pitblado, B.L. 2011. "A Tale of Two Migrations: Reconciling Recent Biological and Archaeological Evidence for the Pleistocene Peopling of the Americas." *Journal of Archaeological Research* 19(4): 327–375.

Pitblado, B.L. 1998. "Peak to Peak in Paleoindian Time: Occupation of Southwest Colorado." *Plains Anthropologist* 43(166): 333–348.

Potter, B.A., Irish, J.D., Reuther, J.D., et al. 2011. "A Terminal Pleistocene

Child Cremation and Residential Structure from Eastern Beringia." *Science* 331: 1058–1062.

———. 2014. "New Insights into Eastern Beringian Mortuary Behavior: A Terminal Pleistocene Double Infant Burial at Upward Sun River." *Proceedings of the National Academy of Sciences* 111(48): 17060–17065.

Powell Adam, et al. 2009. "Late Pleistocene Demography and the Appearance of Modern Human Behavior." *Science* 324: 1298.

Prasciunas, M.M., 2011. "Mapping Clovis: Projectile Points, Behavior, and Bias." *American Antiquity* 76(1): 107–126.

Purdy B.A., Jones, K.S., et al. 2011. "Earliest Art in the Americas: Incised Image of a Proboscidean on a Mineralized Extinct Animal Bone from Vero Beach, Florida." *Journal of Archaeological Science* 38: 2908–2913.

Redmond, B.G., McDonald, H.G., Greenfield, H.J., et al. 2013. "New Evidence for Late Pleistocene Human Exploitation of Jefferson's Ground Sloth (*Megalonyx jeffersonii*) from Northern Ohio, USA." *World Archaeology* 44(1): 75–101.

Richards, M.P., Jacobi, R., Cook, J., et al. 2005. "Isotope Evidence for the Intensive Use of Marine Foods by Late Upper Palaeolithic Humans." *Journal of Human Evolution* 49: 390–394.

Rifkin, R.F. 2011. "Assessing the Efficacy of Red Ochre as a Prehistoric Hide Tanning Ingredient." *Journal of African Archaeology* 9(2): 131–158.

Rivals, F., Mihlbachler, M.C., Solounias, N. 2010. "Palaeoecology of the Mammoth Steppe Fauna from the Late Pleistocene of the North Sea and Alaska: Separating Species Preferences from Geographic Influence in Paleoecological Dental Wear Analysis." *Palaeogeography, Palaeoclimatology, Palaeoecology* 286: 42–54.

Robinson, B.S., Ort, J.C., Eldridge, W.A., et al. 2009. "Paleoindian Aggregation and Social Context at Bull Brook." *American Antiquity* 74(3): 423–447.

Rothschild, B.M., Laub, R. 2006. "Hyperdisease in the Late Pleistocene: Validation of an Early 20th Century Hypothesis." *Naturwissenschaften* 93: 557–564.

Salazar, D., Jackson, D., et al. 2011. "Early Evidence (ca. 12,000 BP) for Iron Oxide Mining on the Pacific Coast of South America." *Current Anthropology* 52(3): 463–475.

Sanchez, G, Holliday, V.T., Gaines, E., et al. 2014. "Human (Clovis)–Gomphothere (*Cuvieronius* sp.) association ~13,390 Calibrated yBP in Sonora, Mexico." *Proceedings of the National Academy of Sciences* 111(30): 10972–10977.

Schubert, B.W. 2010. "Late Quaternary Chronology and Extinction of North American Giant Short-faced Bears (*Arctodus simus*)." *Quaternary International* 217(1–2): 188–194.

Schubert, B.W., Hulbert Jr., R.C., MacFadden, B.F., et al. 2010. "Giant Short-faced Bears (*Arctodus simus*) in Pleistocene Florida, USA: A Substantial Range Extension." *Journal of Paleontology* 84(1): 79–87.

Schubert, B.W., Wallace, S.C. 2009. "Late Pleistocene Giant Short-faced Bears, Mammoths, and Large Carcass Scavenging in the Saltville Valley of Virginia, USA." *Boreas* 38: 482–492.

Schurr, T., Dulik, M.C., et al. 2012. "Clan, Language, and Migration History Has Shaped Genetic Diversity in Haida and Tlingit Populations from Southeast Alaska." *American Journal of Physical Anthropology* 148: 422–435.

Scott, E. 2010. "Extinctions, Scenarios, and Assumptions: Changes in Latest Pleistocene Large Herbivore Abundance and Distribution in Western North America." *Quaternary International* 217: 225–239.

Semken Jr., H.A., Graham, R.W., Stafford Jr., T.W. 2010. "AMS 14C Analysis of Late Pleistocene Non-analog Faunal Components from 21 Cave Deposits in Southeastern North America." *Quaternary International* 217: 240–255.

Shapiro, B., Cooper, A. 2003. "Beringia as an Ice Age Genetic Museum." *Quaternary Research* 60: 94–100.

Sicoli, M.A., Holton, G. 2014. "Linguistic Phylogenies Support Back-migration from Beringia to Asia." *PLoS ONE* 9(3): e91722. doi:10.1371/journal.pone.0091722.

Siebe, C., Schaaf, P., Urrutia-Fucugauchi, J., et al. 1999. "Mammoth Bones Embedded in a Late Pleistocene Lahar from Popocatépetl Volcano, Near Tocuila, Central México." *Geological Society of America Bulletin* 111(10): 1550–1562.

Smith, G.M. 2010. "Footprints Across the Black Rock: Temporal Variability in Prehistoric Foraging Territories and Toolstone Procurement Strategies in the Western Great Basin." *American Antiquity* 75(4): 865–885.

Smith, P.E.L. 1962. "Solutrean Origins and the Question of Eastern diffusion." *Arctic Anthropology* 1(1): 58–67.

Speth, J.D., Newlander, K., White, A.A., et al. 2013. "Early Paleoindian Big-Game Hunting in North America: Provisioning or Politics?" *Quaternary International*, 285: 111–139.

Stafford, M.D., Frison, G.C., et al. 2003. "Digging for the Color of Life: Paleoindian Red Ochre Mining at the Powars II Site, Platte County, Wyoming, U.S.A." *Geoarchaeology* 18(1): 71–90.

Stanford, D.J., Bradley, B.A. 2012. *Across Atlantic Ice: The Origin of America's Clovis Culture*. (Berkeley: University of California Press).

Steele, J. 2010. "Radiocarbon Dates as Data: Quantitative Strategies for Estimating Colonization Front Speeds and Event Densities." *Journal of Archaeological Science* 37: 2017–2030.

Steffen, M.L. and Harington, C.R. 2010. "Giant Short-Faced Bear (*Arctodus simus*) from Late Wisconsinan Deposits at Cowichan Head, Vancouver Island, British Columbia." *Canadian Journal of Earth Sciences* 47: 1029–1036.

Stiger, M. 2006. "A Folsom Structure in the Colorado Mountains." *American Antiquity* 71(2): 321–351.

Stout, B. 1986. "Discovery and C14 Dating of the Black Rock Desert Mammoth." *Nevada Archaeologist* 5(2): 21–23.

Straus, L. 2011. "Humans and Younger Dryas: Dead End, Short Detour, or Open Road to the Holocene?" *Quaternary International* 242: 259–261.

Straus, L.G., Meltzer, D.J, and Goebel, T. 2005. "Ice Age Atlantis? Exploring the Solutrean-Clovis 'Connection.'" *World Archaeology* 37(4): 507–532.

Strong, W.D. 1934. "North American Indian Traditions Suggesting a Knowledge of the Mammoth." *American Anthropologist New Series* 36: 81–88.

Surovell, T.A. 2003. "Simulating Coastal Migration in New World Colonization." *Current Anthropology* 44(4): 580–591.

———. 2000. "Early Paleoindian Women, Children, Mobility, and Fertility." *American Antiquity* 65(3): 493–508.

Surovell, T.A., Waguespack, N.M. 2008. "How Many Elephant Kills Are 14? Clovis Mammoth and Mastodon Kills in Context." *Quaternary International* 191: 82–97.

Surovell, T.A., Waguespack, N.M., Brantingham, P.J. 2005. "Archaeological Evidence for Proboscidean Overkill." *Proceedings of the National Academy of Sciences* 102: 6231–6236.

Tackney, J.C., Potter, B.A., Raff, J., et al. 2015. "Two Contemporaneous Mitogenomes from Terminal Pleistocene Burials in Eastern Beringia." *Proceedings of the National Academy of Sciences* 112(45): 13833–13838.

Tankersley, K.B., Tankersley, K.O., Shaffer, N.R., et al. 1995. "They Have a Rock That Bleeds: Sunrise Red Ochre and Its Early Paleoindian Occurrence at the Hell Gap Site, Wyoming." *Plains Anthropologist* 40(152): 185–194.

Teale, C.L., Miller, N.G. 2012. "Mastodon Herbivory in Mid-latitude Late-Pleistocene Boreal Forests of Eastern North America." *Quaternary Research* 78: 72–81.

Teit, J.A., 1917. "Kaska Tales." *The Journal of American Folklore* 30(118): 427–473.

Thomas, M.G. 2012. "The Flickering Genes of the Last Mammoths." *Molecular Ecology* 21: 3379–3381.

Van Valkenburgh, B., Hertel, F. 1993. "Tough Times at La Brea: Tooth

Breakage in Large Carnivores of the Late Pleistocene." *Science* 261(5120): 456–459.

Vygotsky, L.S. 1978. *Mind in Society: The Development of Higher Psychological Processes.* Cambridge: Harvard University Press.

Waguespack, N.M. 2005. "The Organization of Male and Female Labor in Foraging Societies: Implications for Early Paleoindian Archaeology." *American Anthropologist* 107(4): 666–676.

Walker, D.A., Bockheim, J.G., Chapin III, F.S., et al. 2001. "Calcium-rich Tundra, Wildlife, and the 'Mammoth Steppe.'" *Quaternary Science Reviews* 20(149): 149–163.

Wang, S., Lewis Jr., C.M., et al. 2007. "Genetic Variation and Population Structure in Native Americans." *PLoS Genetics* 3(11): e185. doi:10.1371/journal .pgen.0030185.

Waters, M.R. 2011. "The Buttermilk Creek Complex and the Origins of Clovis at the Debra L. Friedkin Site, Texas." *Science* 331: 1599–1603.

Waters, M.R., Forman, S.L., Stafford Jr. T.W., et al. 2009. "Geoarchaeological Investigations at the Topper and Big Pine Tree sites, Allendale County, South Carolina." *Journal of Archaeological Science* 36(7): 1300–1311.

Waters, M.R., Stafford Jr., T.W., McDonald, H.G., et al. 2011. "Pre-Clovis mastodon hunting 13,800 years ago at the Manis Site, Washington." *Science*, 334: 351–353.

Wells P.V., Stewart, J.D. 1987. "Cordilleran-boreal Taiga and Fauna on the Central Great Plains of North America, 14,000–18,000 years ago." *The American Midland Naturalist* 118(1): 94–106.

Wilson, M., Kenady, S., et al. 2009. "Late Pleistocene *Bison antiquus* from Orcas Island, Washington, and the Biogeographic Importance of an Early Postglacial Land Mammal Dispersal Corridor from the Mainland to Vancouver Island." *Quaternary Research* 71(1), 49–61.

Woodman, N., Athfield, N.B. 2009. "Post-Clovis Survival of American Mastodon in the Southern Great Lakes Region of North America." *Quaternary Research* 72: 359–363.

Wreschner, E.E., Bolton, R., Butzer, K.W., et al. 1980. "Red Ochre and Human Evolution: A Case for Discussion." *Current Anthropology* 21(5): 631–644.

Yesner, D.R. 2001. "Human Dispersal into Interior Alaska: Antecedent Conditions, Mode of Colonization, and Adaptations." *Quaternary Science Reviews* 20: 315–327.

Yi, M., Barton, L., Morgan, C., et al. 2013. "Microblade Technology and the Rise of Serial Specialists in North-central China." *Journal of Anthropological Archaeology* 32: 212–223.

Yulea, J.V., Jensen, C.X.J., Joseph, A. 2009. "The Puzzle of North America's Late Pleistocene Megafaunal Extinction Patterns: Test of New Explanation Yields Unexpected Results." *Ecological Modelling* 220: 533–544.

Zazulaa, G.D., Schwegera, C.E., Beaudoin, A.B., et al. 2006. "Macrofossil and Pollen Evidence for Full-glacial Steppe Within an Ecological Mosaic Along the Bluefish River, Eastern Beringia." *Quaternary International* 142–143: 2–19.

INDEX

Page numbers in *italics* refer to illustrations.

St. Lawrence Island,
Alaska ✳

Yukon Flats,
Alaska ✳   Old Crow Basin,
                           Yukon ✳

          Swan Point,   ✳ Bluefish Caves,
          Alaska ✳   Yukon

Harding Icefield,
Alaska ✳ ✳ Prince William Sound,
            Alaska

          On Your Knees Cave, ✳
          Alaska

          Triquet Island,   Wally's Beach
          British Columbia ✳   Albert

          Calvert Island,
          British Columbia

                                   Wilson Butte Cave,
                                   Idaho

                           Paisley Caves, ✳
                           Oregon

                           Black Rock Desert, ✳
                           Nevada

                                   Channel Islands,
                                   California